Marko Dragicevic

Silicon Strip Detectors for High Luminosity Hadron Colliders

Marko Dragicevic

Silicon Strip Detectors for High Luminosity Hadron Colliders

Basics on silicon strip detector design and production and new developments for the upgrade of CMS Tracker

Südwestdeutscher Verlag für Hochschulschriften

Impressum/Imprint (nur für Deutschland/ only for Germany)
Bibliografische Information der Deutschen Nationalbibliothek: Die Deutsche Nationalbibliothek verzeichnet diese Publikation in der Deutschen Nationalbibliografie; detaillierte bibliografische Daten sind im Internet über http://dnb.d-nb.de abrufbar.

Alle in diesem Buch genannten Marken und Produktnamen unterliegen warenzeichen-, marken- oder patentrechtlichem Schutz bzw. sind Warenzeichen oder eingetragene Warenzeichen der jeweiligen Inhaber. Die Wiedergabe von Marken, Produktnamen, Gebrauchsnamen, Handelsnamen, Warenbezeichnungen u.s.w. in diesem Werk berechtigt auch ohne besondere Kennzeichnung nicht zu der Annahme, dass solche Namen im Sinne der Warenzeichen- und Markenschutzgesetzgebung als frei zu betrachten wären und daher von jedermann benutzt werden dürften.

Verlag: Südwestdeutscher Verlag für Hochschulschriften Aktiengesellschaft & Co. KG
Dudweiler Landstr. 99, 66123 Saarbrücken, Deutschland
Telefon +49 681 37 20 271-1, Telefax +49 681 37 20 271-0
Email: info@svh-verlag.de
Zugl.: Wien, TU, Diss., 2010

Herstellung in Deutschland:
Schaltungsdienst Lange o.H.G., Berlin
Books on Demand GmbH, Norderstedt
Reha GmbH, Saarbrücken
Amazon Distribution GmbH, Leipzig
ISBN: 978-3-8381-1894-9

Imprint (only for USA, GB)
Bibliographic information published by the Deutsche Nationalbibliothek: The Deutsche Nationalbibliothek lists this publication in the Deutsche Nationalbibliografie; detailed bibliographic data are available in the Internet at http://dnb.d-nb.de.

Any brand names and product names mentioned in this book are subject to trademark, brand or patent protection and are trademarks or registered trademarks of their respective holders. The use of brand names, product names, common names, trade names, product descriptions etc. even without a particular marking in this works is in no way to be construed to mean that such names may be regarded as unrestricted in respect of trademark and brand protection legislation and could thus be used by anyone.

Publisher: Südwestdeutscher Verlag für Hochschulschriften Aktiengesellschaft & Co. KG
Dudweiler Landstr. 99, 66123 Saarbrücken, Germany
Phone +49 681 37 20 271-1, Fax +49 681 37 20 271-0
Email: info@svh-verlag.de

Printed in the U.S.A.
Printed in the U.K. by (see last page)
ISBN: 978-3-8381-1894-9

Copyright © 2010 by the author and Südwestdeutscher Verlag für Hochschulschriften Aktiengesellschaft & Co. KG and licensors
All rights reserved. Saarbrücken 2010

Abstract

To facilitate the understanding of the basic principles of silicon strip sensors, the first part of this book reviews the fundamentals in semiconductor technology. The necessary process steps to manufacture strip sensors in a so-called planar process are described in detail and the effects of irradiation on silicon strip sensors are discussed.

To conclude the introductory part of the book, the CMS experiment is described as a living example of a high energy experiment using silicon strip sensors as a tracking detector. The CMS Tracker is described in detail where the choice of the substrate material and the complex geometry of the sensors are reviewed and the quality assurance procedures for the production of the sensors are presented.

The main part of the book deals with the demands of a high luminosity collider in the example of the proposed upgrade of the LHC accelerator (sLHC). Chapter 5 discusses the proposed upgrade scenario for the sLHC and and summarises the main challenges posed on the CMS Tracker. The subsequent chapter reviews the latest results on radiation hard sensor technologies gathered from CERN's RD50 collaboration. Advancements to the quality assurance techniques used for the CMS Tracker are described in chapter 7 while the last chapter concludes the book with an in-depth description of the development of a new sensor-to-readout connection technique.

Contents

I. Theoretical and Technical Background Information   1

1. Basics on Silicon Semiconductor Technology   5
    1.1. Intrinsic Properties of Silicon 6
    1.2. Extrinsic Properties of Doped Silicon 10
    1.3. Carrier Transport 15
       1.3.1. Drift 15
       1.3.2. Diffusion 15
    1.4. Carrier Generation and Recombination 16
       1.4.1. Thermal Generation 17
       1.4.2. Generation by Electromagnetic Excitation 17
       1.4.3. Generation by Charged Particles 17
       1.4.4. Charge Carrier Lifetime 20
    1.5. Basic Semiconductor Structures 22
       1.5.1. The p-n Junction or Diode 22
       1.5.2. The n+-n or p+-p Junction 29
       1.5.3. The Metal - Semiconductor Contact 29
       1.5.4. The Metal - Oxide - Semiconductor Structure 30
       1.5.5. The Polysilicon Resistor 37
    1.6. Radiation Damage and the NIEL hypothesis 38
       1.6.1. Bulk and Surface Damage 39
       1.6.2. Changes in Properties due to Defect Complexes 41
       1.6.3. Annealing 43
       1.6.4. Reverse Annealing 44

2. Silicon Strip Sensors   47

Contents

    2.1. Working Principle 48
    2.2. Design Basics of a Silicon Strip Sensor 51
        2.2.1. Strip Geometry 52
        2.2.2. DC to AC coupled Strips 53
        2.2.3. Biasing of the Strips 53
        2.2.4. Breakdown Protection 55
        2.2.5. Contact Pads 57
    2.3. Manufacturing of Silicon Sensors 59
        2.3.1. Silicon for Silicon Sensors 59
        2.3.2. General Process Steps in Planar Technology 61
        2.3.3. A Showcase Process Sequence 66

3. The CMS Experiment   75

    3.1. Tracking System 79
        3.1.1. Pixel Tracker 79
        3.1.2. Strip Tracker 79
    3.2. Calorimeter System 82
        3.2.1. Electromagnetic Calorimeter 82
        3.2.2. Hadronic Calorimeter 83
    3.3. Muon System 84
    3.4. Trigger System 85

4. The CMS Silicon Strip Sensors   87

    4.1. Sensor Design 88
        4.1.1. Choice of Bulk Material 88
        4.1.2. Sensor Geometry 89
        4.1.3. Strip Geometry 91
    4.2. Sensor Quality Assurance 91
    4.3. Detector Module Design 92

II. Sensor Design for the new CMS Tracker   95

5. Super LHC and the CMS Upgrade   99

    5.1. Super LHC: a luminosity upgrade 100
    5.2. Challenges for the Strip Tracker 104
        5.2.1. Increase in granularity 104

	5.2.2.	Increase in Radiation 107
	5.2.3.	Tracking Trigger 109
5.3.	Summary	 112

6. New Bulk Materials and Process Technologies 113
6.1. Combined Results from RD50 114
    6.1.1. Full Depletion Voltage and Effective Doping Concentration 115
    6.1.2. Charge Collection Efficiency 119
    6.1.3. Reverse Bias Current 120
    6.1.4. Charge Multiplication in Silicon 121
6.2. Summary 123
6.3. Outlook 124

7. Quality Assurance using Test Structures 125
7.1. Revising and Enhancing the Standard Set of Test Structures 126
7.2. First Production of the Enhanced Set of Test Structures 131
    7.2.1. Results 132
7.3. Second Production of the Enhanced Set of Test Structures 135
    7.3.1. Results 137
7.4. Summary and Outlook 141

8. Sensors with Integrated Pitch Adapters 143
8.1. On-Sensor Integration 145
8.2. Design of the Prototype Sensors 146
    8.2.1. Main Strip Parameters 147
    8.2.2. Pitch Adapter Geometry 147
    8.2.3. Bump Bonding 148
    8.2.4. Large Sensor with 512 Strips 148
8.3. Structure Summary and Wafer Layout 152
8.4. Electrical Characterisation 158
    8.4.1. Process Quality Control 159
    8.4.2. Sensor Quality Control 160
8.5. Beam Tests at CERN's SPS 166
    8.5.1. Module Construction 166
    8.5.2. Readout Electronics and Services 169
    8.5.3. The Beam 169

		8.5.4.	The Test Setup 169
		8.5.5.	Results 171
	8.6.	Summary 181

9. Conclusion and Outlook   **183**

Bibliography   **185**

Part I.

Theoretical and Technical Background Information

Introduction

To put the developments in silicon strip technology into perspective, it is important to review the basic principles of semiconductor theory and the current technological status of silicon strip sensors.

The first part of this book reviews the background on semiconductor theory. The relevant effects necessary to understand the working principle of silicon strip sensor are presented including the effects that degrade the performance of such a device due to irradiation. The second chapter reviews the basic working principle of a strip sensor and subsequently summarises the most important design features of a modern strip sensor and their manufacturing process.

The last two chapter provide an overview of the CMS Experiment followed by an in-depth description of the sensors and detector modules designed for and operated within the tracking system of the CMS experiment.

*Silicon, (Latin: silicium) is the most common metalloid. It is a chemical element, which has the symbol **Si** and atomic number 14. The atomic mass is 28.0855. A tetravalent metalloid, silicon is less reactive than its chemical analog carbon. As the eighth most common element in the universe by mass, silicon very rarely occurs as the pure free element in nature, but is more widely distributed in dusts, planetoids and planets as various forms of silicon dioxide (silica) or silicates. On Earth, silicon is the second most abundant element (after oxygen) in the crust, making up 25.7% of the crust by mass.*

Wikipedia on **Silicon**

1
Basics on Silicon Semiconductor Technology

Most modern high energy physics experiments use silicon strip sensors for tracking the trajectories of charged particles. Such detectors can be built with a very moderate material budget, offering precise spatial resolution and robust performance even after high irradiation. Costs can be kept within reasonable amounts, as the production technology of the sensors follows common processes developed in the IC industry. The following chapters review the basic fundamentals needed to understand the operation of such sensors and how they are manufactured.

1.1. Intrinsic Properties of Silicon

Silicon is the 14th element in the periodic table of elements and belongs to group IV. It has four covalently bound electrons and forms a face-centered cubic structure with a lattice spacing of 5.430710 Å (0.5430710 nm) which is shown in figure 1.1. The most important properties of silicon are noted in table 1.1. To comprehend the chemical and electronic properties of solid silicon, it is important to understand the energy band structure of solids as seen in figure 1.2

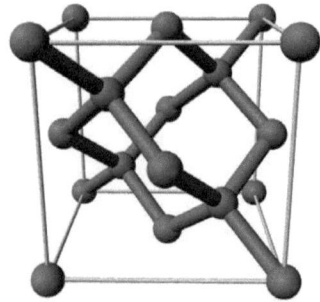

Figure 1.1.: The face-centered cubic structure of silicon which is the same as for diamond.

1.1. Intrinsic Properties of Silicon

Substance	Si	Ge	GeAs	Diamond
lattice	diamond	diamond	zinc blende	diamond
atomic number	5.4307	5.657	5.653	3.5668
average atomic mass	14	32	31+33	6
density [g/cm^3]	2.329	5.323	5.317	3.515
melting point [°C]	1415	937	1238	3907
thermal expansion coefficient [10^{-6}/K]	2.56	5.90	6.86	1.0
thermal conductivity [W/cmK]	1.56	0.60	-.45	10
intrinsic resistivity	230 kΩ cm	47 Ω cm		
radiation length [cm]	9.36	2.30		12.15
refractive index	3.42	3.99	3.25	2.42
dielectric constant	11.9	16.2	12.9	5.7
breakdown field [V/cm]	$\approx 3 \times 10^5$			
band gap	1.12	0.67	1.42	5.48
intrinsic carrier concentration [1/cm$_3$]	1.45×10^{10}	2.33×10^{13}	2.1×10^6	
mean energy for e$^-$/h$^+$ creation [eV]	3.63	2.96	4.35	13.1
drift mobility electrons [cm^2/Vs]	1450	3900	8800	1800
drift mobility holes [cm^2/Vs]	505	1800	320	1600

Table 1.1.: Important properties of silicon and other commonly used semiconductor material[1].

The *valence band* is the highest band filled with electrons at absolute zero temperature with the upper energy border E_V. Electrons from this band are bound to its nucleus and cannot move along the crystal. At a temperature of $T = 0$ K the valence band would be fully occupied. The lowest energy band that is not occupied and would be empty at a temperature of $T = 0$ K is the *conduction band* with its lowest energy border E_C. Electrons in the conduction band can move freely between atoms of the crystal. The forbidden area in between the two bands $E_g = E_c - E_v$ is called *band gap*.

Depending on the size of E_g three types of solids can be distinguished:

$E_g \gg 0$: **Insulator** If the band gap is very large the conduction band is almost completely empty and no charge carriers are available to transport an electric current. Such solids are called insulators.

$E_g > 0$: **Semiconductor** If the band gap is only small, electrons can be easily excited into the conduction band. The conduction band at room temperature is therefore sparsely

1. Basics on Silicon Semiconductor Technology

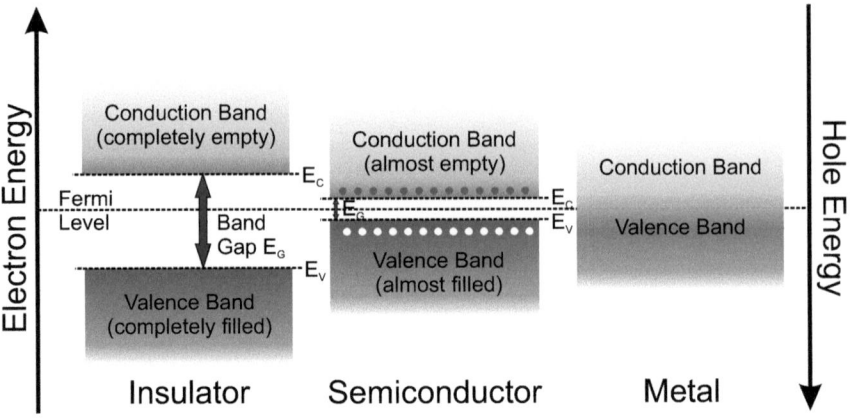

Figure 1.2.: A simplified model of the electron energy band structure in solids.

populated by thermally excited electrons. The maximum size of the band gap, still categorizing a material as semiconductor, is arbitrarily chosen whereas in literature it is usually specified as 5 eV < E_g < 3 eV.

$E_g < 0$: **Metal** If there is no band gap, but the valence and the conduction band are overlapping, the material is called a metal. Enough electrons are available to transport electric currents.

For intrinsic silicon the band gap $E_g^{Si} = 1.12$ eV, therefore silicon is classified as semiconductor.

In a semiconductor, the electrons can transit from the valence to the conduction band through excitation by phonons. Phonons are quasi-particles related to the quantised vibration modes of the crystal lattice. The number of charge carriers (electrons and holes) can be estimated under equilibrium conditions. The occupation probability for an electronic state is given by Fermi-Dirac-statistics:

$$F(E) = \frac{1}{1 + \exp\left(\frac{E - E_{fermi}}{k_B T}\right)} \tag{1.1}$$

where T is the temperature, k_B is he Boltzmann constant and E_{fermi} is the Fermi energy level at which the occupation probability is one half. For electrons with $|E - E_{fermi}| > 3 k_B T$ the Fermi-Dirac function can be approximated by:

$$F_e(E) \approx \exp\left(-\frac{E - E_{fermi}}{k_B T}\right) \tag{1.2}$$

1.1. Intrinsic Properties of Silicon

The same can be done for the non occupied states or holes which can be treated as quasi-particles. With $F_e + F_h = 1$:

$$F_h(E) = 1 - F_e(E) \approx \exp\left(-\frac{E_{fermi} - E}{k_B T}\right) \tag{1.3}$$

The state density near the bottom of the conduction bad is given by:

$$N(E) = \frac{\left(2m_e^{eff}\right)^{\frac{3}{2}}}{2\pi^2 \hbar^3 \sqrt{E - E_c}} \tag{1.4}$$

where the effective electron mass $m_e^{eff} = 0.32\, m_e$ and depends on the orientation of the silicon lattice. Convoluting the state density $N(E)$ with the corresponding occupation probability $F_e(E)$ gives the free electron density n:

$$n = \int_{E_c}^{\infty} N(E)F(E)dE = \frac{2}{\hbar^3}\left(2\pi m_e^{eff} k_B T\right)^{\frac{3}{2}} \exp\left(-\frac{E_c - E_{fermi}}{k_B T}\right) \tag{1.5}$$

$$= N_c \exp\left(-\frac{E_c - E_{fermi}}{k_B T}\right) \tag{1.6}$$

and the free hole density respectively:

$$p = N_v \exp\left(-\frac{E_{fermi} - E_v}{k_B T}\right) \tag{1.7}$$

For an intrinsic semiconductor at thermal equilibrium, the free electron density and the free hole density are equal and is given by:

$$n_i = \sqrt{np} = n = p = \sqrt{N_c N_v} \exp\left(-\frac{E_g}{2k_B T}\right) \tag{1.8}$$

For intrinsic silicon this results in $n_i = n = p \approx 1.45 \times 10^{10}$ cm^{-3} at $T = 300$ K and the Fermi energy level is given by:

$$E_{fermi,i} = \frac{E_c + E_v}{2} + \frac{3k_B T}{4} \ln\left(\frac{m_p}{m_n}\right) \tag{1.9}$$

which is usually located close to the middle of the band gap for small temperatures.

1.2. Extrinsic Properties of Doped Silicon

The intrinsic electric properties of pure silicon as discussed in the previous section, can be manipulated by replacing silicon atoms from the crystal lattice with foreign atoms, so-called dopands. Elements from group III, which have one electron less than silicon or elements from group V, which have an additional electron, can be used to alter the number of charge carriers as seen in figure 1.3. The former elements are then called *acceptors* (as they can accept an additional electron at their place in the crystal lattice) and the later ones are called *donors* (as they can donate an additional electron).

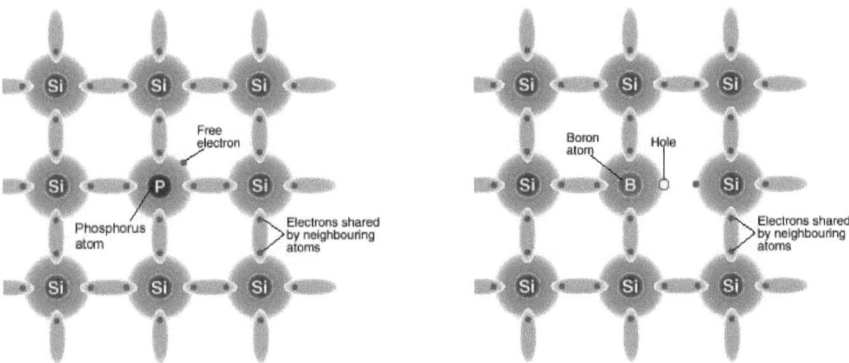

Figure 1.3.: The crystal lattice of silicon doped with phosphorus becomes n-type material (left), and silicon doped with boron becomes a p-type material (right).

The doping of silicon with a donator, usually phosphorus (P) or arsenic (As), leads to an increase of negative charge carriers. These weakly bond valence electrons introduce energy states slightly below the lower conduction band edge E_c as seen in figure 1.4. For phosphorus ($E_C - E_D = 0.045$ eV) or arsenic ($E_C - E_D = 0.054$ eV) in silicon, these states are almost fully ionized at room temperature and the electrons will be pushed into the conduction band due to the many states with similar energy in the conduction band, with which the donor states have to share their electrons. Doped silicon with an excess of negative charge carriers is called *n-type* silicon.

If silicon is doped with an acceptor, usually boron (B), the missing electrons will act like a positive charge, thus increasing the number of positive charge carriers. These holes introduce energy states slightly above the upper valence band energy E_v as seen in figure 1.4. For boron ($E_A - E_V = 0.045$ eV) in silicon, these states are almost filled completely at room temperature

1.2. Extrinsic Properties of Doped Silicon

and holes will be left in the valence band. This type of silicon is also called *p-type* silicon.

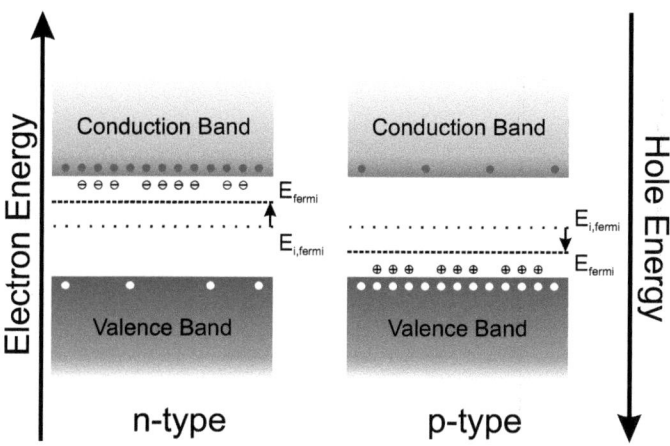

Figure 1.4.: Band structure for doped semiconductors. The n-type material on the left has donor states slightly below the conduction band (left), while for p-type materials, the acceptor states are slightly above the valence band.

The Fermi level E_{fermi} of a doped semiconductor is shifted towards the conduction band in n-type material and towards the valence band in p-type material as seen in figure 1.5. The new Fermi level can be derived from equations 1.6 and 1.7 for n-type and p-type silicon respectively. With the approximation that the electron and hole densities are equal to the number of donator and acceptor atoms or $n = N_D$ for n-type silicon and $p = N_A$ for p-type silicon the new Fermi level gets:

$$E_{fermi,n} = E_c - k_B T \ln\left(\frac{N_c}{N_D}\right), \text{ for n-type} \qquad (1.10)$$

$$E_{fermi,p} = E_v - k_B T \ln\left(\frac{N_v}{N_A}\right), \text{ for p-type} \qquad (1.11)$$

and the charge carrier concentration is:

$$n = n_i \exp\left(-\frac{E_{fermi} - E_i}{k_B T}\right) \qquad (1.12)$$

$$p = n_i \exp\left(-\frac{E_i - E_{fermi}}{k_B T}\right) \qquad (1.13)$$

1. Basics on Silicon Semiconductor Technology

According to the mass action law, an increase in majority charge carriers must be accompanied by a decrease of minority carriers so that $n \cdot p = n_i^2$. In case of n-type material, the majority carriers are electrons while for p-type material they are the minority carriers and the holes are the majority carriers.

The temperature dependence of the charge carrier concentration in n-type silicon is seen in figure 1.6. The extrinsic behaviour of doped silicon is only apparent at medium temperatures, while for high temperatures it becomes intrinsic again. This is caused by the increase of electrons populating the conduction band due to thermal excitation where the intrinsic electron concentration gets similar or higher then the concentration of donor atoms.

The resistivity of an extrinsic semiconductor is defined by the concentration of majority charge carriers. At medium temperatures, within the extrinsic region, this is equal to the doping concentration and we get:

$$\rho = \frac{1}{q\mu_n N_D}, \text{ for } N_D \ll N_A \tag{1.14}$$

$$\rho = \frac{1}{q\mu_p N_A}, \text{ for } N_A \ll N_D \tag{1.15}$$

1.2. Extrinsic Properties of Doped Silicon

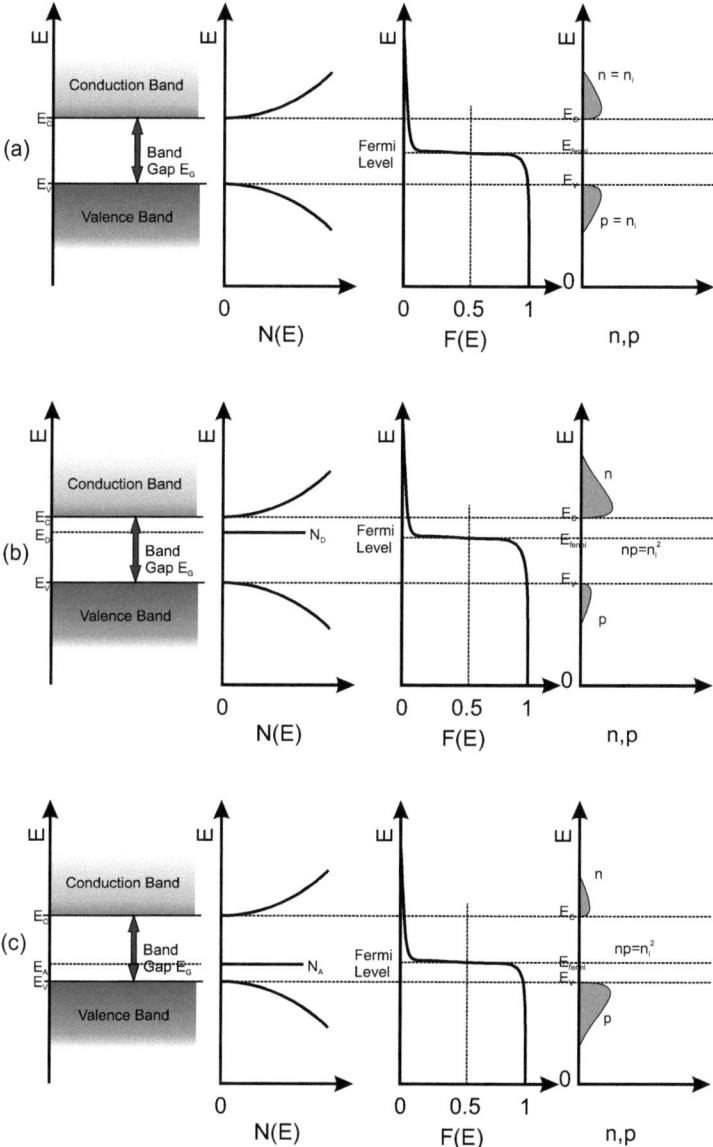

Figure 1.5.: State density, Fermi levels and free charge density in silicon. The density of states $N(E)$ is the same for intrinsic (a) and doped silicon (b and c). The Fermi levels E_{fermi} for n-type (b) and p-type (c) silicon and the final free charge densities change.

1. Basics on Silicon Semiconductor Technology

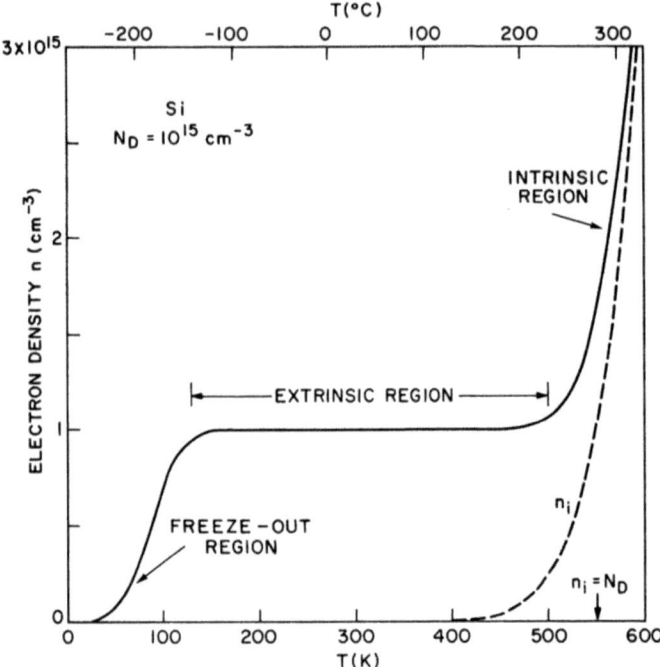

Figure 1.6.: Electron density in n-type silicon as function of temperature [2]. At low temperature the thermal energy is to low to ionise the donator atoms. For medium temperatures the concentration of electrons in the conduction band is equal to the number of donators atoms. At high temperatures ($k_B T \approx E_G$) the intrinsic electron concentration is similar to the number of donator atoms and the doped semiconductor shows intrinsic behaviour.

1.3. Carrier Transport

In the previous sections we discussed semiconductors in equilibrium without any external fields applied to the material and homogeneous distribution of charge carriers. The following sections will review the phenomena occurring in semiconductors in a non-equilibrium state.

Electrons in the conduction band and holes in the valence band can move freely within the semiconductor as they are not bound to a specific lattice site. Their mean kinetic energy is $3/2 k_B T$ which gives them a velocity of about 10^7 cm/s at room temperature. The free charge carriers can scatter on imperfections of the crystal lattice due to thermal vibrations, on impurities and on defects. The mean free path is about 10^{-5} cm with a mean free time of $\tau_c \approx 10^{-12}$ s.

1.3.1. Drift

The average displacement of a charge carrier in the field free case is zero. If an electrical field E is applied, the carriers will drift according to their charge and the field orientation. If the acceleration between two random collisions is small enough that the resulting velocity change is small compared to the thermal energy, the average drift velocity can be calculated by:

$$v_n = -\frac{q\tau_c}{m_n}E = -\mu_n E \quad (1.16)$$

$$v_p = \frac{q\tau_c}{m_p}E = \mu_p E \quad (1.17)$$

We can now calculate the current density:

$$J_n = -qnv_n = qn\mu_n E \quad (1.18)$$

$$J_p = qnv_p = qn\mu_p E \quad (1.19)$$

If the field is high enough, strong deviations from linearity can be observed and the drift velocities finally become independent of the applied electric field with saturation at $\mu_{n,s}$ and $\mu_{p,s}$.

The carrier mobilities μ_n and μ_p depend on temperature and doping concentration due to the scattering mechanism on imperfections and defects of the crystal lattice and on impurities like doping atoms. For a more detailed description see [2].

1.3.2. Diffusion

Diffusion describes the transport of particles from a region of higher concentration to a region of lower concentration by random particle movement. To describe the process of charge carrier diffusion in a semiconductor, we consider an inhomogeneous distribution of charge carriers while we neglect any effects due to electric fields, even the fields created due to the doping atoms and the inhomogeneous distribution of charge carriers themselves. Basically electrons and holes are considered neutral.

The movement of charge carriers can then be described by the diffusion equation:

$$F_n = -D_n \nabla n \tag{1.20}$$

$$F_p = -D_p \nabla p \tag{1.21}$$

where F_n (F_p) is the flux of carriers and D_n (D_p) is the diffusion constant.

By combining equations 1.18, 1.19 with 1.20, 1.21, we obtain the current densities for drift and diffusion:

$$J_n = qn\mu_n \varepsilon + qD_n \nabla n \tag{1.22}$$

$$J_p = qp\mu_p \varepsilon - qD_p \nabla p \tag{1.23}$$

To derive Einsteins equation from equations 1.22 and 1.23, we consider a system at equilibrium where the current densities have to be zero at any point of the system and get:

$$D_n = \frac{k_B T}{q} \mu_n \tag{1.24}$$

$$D_p = \frac{k_B T}{q} \mu_p \tag{1.25}$$

1.4. Carrier Generation and Recombination

Free charge carriers are generated by lifting electrons from the valence band up to the conduction band, which creates a free electron and a free hole simultaneously. The necessary energy for the excitation can be acquired from a number of mechanisms like thermal agitation, optical excitation or ionisation by charged particles.

Another possibility to influence the number of charge carriers is the injection of charge using a forward biased diode or deplete the number of charge carriers using a reverse bias diode. Basic semiconductor structures like diodes will be discussed in section 1.5 while the mechanisms for charge carrier generation and recombination will be discussed in the following sections.

1.4.1. Thermal Generation

Thermally generated charge carriers are to be considered as a noise source for semiconductor sensors. If the band gap is small enough in relation to the thermal voltage at room temperature ($k_BT/q = 0.0259$ V for $T = 300$ K) electrons get excited into the conduction band creating a electron - hole pair. In a sensor, this randomly created charge would overlay the charge created by the signal to be measured.

For silicon, the band gap is large and therefore the probability of direct excitation is low at room temperature. The excitation usually occurs by intermediate states within the band gap which are created by imperfections and impurities inside the crystal lattice. For indirect semiconductors such as silicon, the required energy for the transfer is not only given by the size of the band gap. Additional energy has to be supplied as the maximum energy in the valence band and the minimum energy in the conduction band are located at different local momenta. For silicon the energy required to lift a valence electron into the conduction band is 3.6 eV.

1.4.2. Generation by Electromagnetic Excitation

Electrons can be excited from the valence into the conduction band by absorbing a photon with an energy larger than the band gap. This effect is used in photo detectors and solar cells.

If the energy supplied by the photon is larger than the band gap, the electron and the hole will gradually move towards the band gap edges by emitting the excess energy into the crystal lattice as phonons. The absorption of energies smaller than the band gap is possible if intermediate states inside the band gap are created by impurities or imperfections of the crystal lattice.

1.4.3. Generation by Charged Particles

Charged particles traversing a material lose part of their kinetic energy by electromagnetic interactions with the electrons and nuclei of the material. This effect is different for electrons and positrons compared to heavier particles.

Electrons and positrons

High energy electrons and positrons predominantly lose their kinetic energy by bremsstrahlung while at low energies the main loss is due to ionisation, although other effects like Møller and Bhabha scattering contribute as well. The individual contributions for different incident energies can be seen in figure 1.7

1. Basics on Silicon Semiconductor Technology

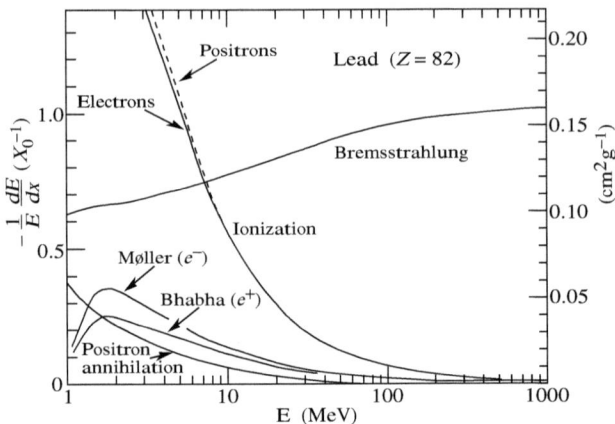

Figure 1.7.: Fractional energy loss per radiation length in lead as a function of electron or positron energy [3]. Electron (positron) scattering is considered as ionization when the energy loss per collision is below 0.255 MeV, and as Møller (Bhabha) scattering when it is above.

The overall energy loss of high energy electrons and positrons can be characterised by the radiation length X_0, which describes the average amount of matter that is traversed by the particles while loosing $1/e$ of its kinetic energy by bremsstrahlung. According to [3] a compact fit to data yields:

$$X_0 = \frac{716.4 \cdot A}{Z(Z+1)\ln\left(\frac{287}{\sqrt{Z}}\right)} \mathrm{g\ cm^{-2}} \qquad (1.26)$$

where the atomic number $Z > 4$ and A is the mass number. For silicon with a density of $\rho = 2.33$ g/cm^3 the radiation length according to equation 1.26 is $X_0 = 9.36$ g/cm^{-2}.

Other charged particles

The mean rate of energy loss (or stopping power) of particles other than electrons and positrons is described by the famous Bethe-Bloch equation [3]:

$$-\frac{dE}{dx} = Kz^2 \frac{Z}{A}\frac{1}{\beta^2}\left[\frac{1}{2}\ln\frac{2m_ec^2\beta^2\gamma^2 T_{max}}{I^2} - \beta^2 - \frac{\delta(\beta\gamma)}{2}\right] \qquad (1.27)$$

where the symbols represent

1.4. Carrier Generation and Recombination

Symbol	Definition	Units or Value
Z	atomic number of absorber	
A	atomic mass of absorber	
K	$4\pi N_A r_e^2 m_e c^2 / A$	0.307075 MeV cm^{-1}
ze	charge of incident particle	
$m_e c^2$	electron mass $\times c^2$	0.511 keV
β	$\frac{v}{c}$	
γ	$(1-\beta^2)^{-1/2}$	
T	kinetic energy	MeV
I	mean excitation energy	eV
$\delta(\beta\gamma)$	density effect correction	

and T_{max} is the maximum kinetic energy which can be imparted to a free electron in a single collision. Figure 1.8 shows the stopping power for positive muons in copper for a wide range of energies. The minimum ionization loss of a muon is located at approximately 350 MeV. For other particles the minimum ionisation energy is different, e.g. for pions it is around 470 MeV and for protons at around 3.2 GeV. Such particles with an energy located at the minimum are called <u>M</u>inimum <u>I</u>onising <u>P</u>article (**MIP**). In high energy physics, most particles have mean energy loss rates close to the minimum and can be considers as MIPs.

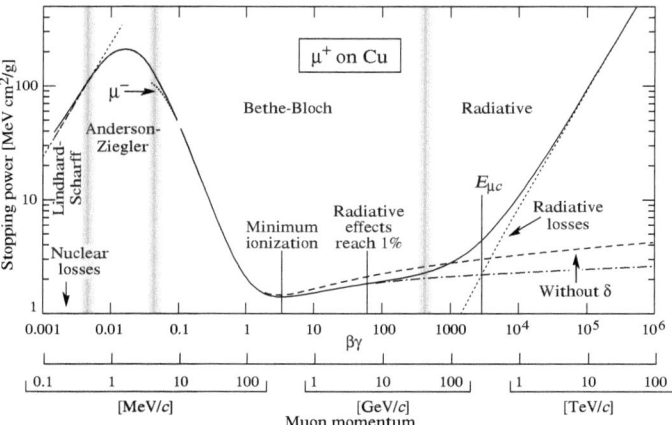

Figure 1.8.: Stopping power ($\langle -dE/dx \rangle$) for positive muons in copper as a function of $\beta\gamma = p/mc$ over nine orders of magnitude in momentum (12 orders of magnitude in kinetic energy) [3]. Solid curves indicate the total stopping power. Vertical bands indicate boundaries between different approximations.

19

For a particle detector, MIPs are an important benchmark, as they deposit only a minimum amount of their kinetic energy in the active sensor material. To assess if a sensor is still able to reliably detect a MIP, the ratio of the induced signal to the noise of the sensor is an important performance parameter. Therefore, the <u>S</u>ignal-to-<u>N</u>oise <u>R</u>atio (**SNR**), defined as the ratio of the signal created by a single particle to the intrinsic noise of the sensor, caused by MIP-like particles is the most relevant description of the sensitivity of a sensor.

Equation 1.27 describes only the *mean* energy loss of a particle in matter, while the actual energy loss of each particle is fluctuating. The statistics of the measurable signal caused by charged particles is described by the Landau distribution as further discussed in chapter 2.1.

1.4.4. Charge Carrier Lifetime

The lifetime of charge carriers in a semiconductor can be described by two parameters, the generation and the the recombination lifetime. They describe the transient behaviour from a state of non-equilibrium, created either due to removal of charge carriers (creation lifetime) or due to injection of additional charge carrier (recombination lifetime), back to equilibrium.

An excess of minority charge carriers can be created by exposing a semiconductor to a light pulse or to ionising radiation. After this initial event creating the additional charge carriers, it will take some time to settle back into equilibrium, where the excess minority charge carriers will recombine with the majority carriers. We assume the overall recombination rate R to be proportional to the product of electron and hole concentration:

$$R = \beta n p = \beta n_i^2 \tag{1.28}$$

A charged particle or a light pulse would cause an additional generation rate G_{rad}. As electrons and holes are created in pairs, it would increase the thermal equilibrium concentration of the minority (n_0) and majority (p_0) charge carriers by the same amount $\Delta n = \Delta p = \Delta$. The recombination rate will increase:

$$R = \beta (n_0 + \Delta)(p_0 + \Delta). \tag{1.29}$$

For a direct semiconductor at thermal equilibrium, the average concentration of holes and electrons is constant in time. Nevertheless the thermal creation (G_{th}) and recombination (R_{th}) of charge carriers is occurring continuously but the rates for both processes are equal:

$$G_{th} = R_{th} = \beta n_0 p_0. \tag{1.30}$$

1.4. Carrier Generation and Recombination

We can now define an excess recombination rate U which is zero at thermal equilibrium:

$$U = R - R_{th} = \beta(np - n_0 p_0). \tag{1.31}$$

The additional generation rate created by ionising radiation G_{rad} must be compensated by excess recombination rate U, thus giving:

$$G_{rad} = U = R - R_{th} = \beta(n_0 + \Delta)(p_0 + \Delta) - \beta n_0 p_0 \tag{1.32}$$
$$= \beta(n_0 p_0 + n_0 \Delta + p_0 \Delta + \Delta^2 - n_0 p_0) \tag{1.33}$$
$$= \beta \Delta (n_0 + p_0 + \Delta) \tag{1.34}$$

For low injection levels, were the number of additional charge carriers is small compared to the number of majority carriers ($\Delta n \ll p_0$ for p-type and $\Delta p \ll n_0$ for n-type material), equation 1.34 simplifies to:

$$G_{rad} = \beta p_0 \Delta n = \frac{\Delta n}{\tau_r} \quad \text{for p-type material with} \quad \tau_r = \frac{1}{\beta p_0} \tag{1.35}$$

$$G_{rad} = \beta n_0 \Delta p = \frac{\Delta p}{\tau_r} \quad \text{for n-type material with} \quad \tau_r = \frac{1}{\beta n_0} \tag{1.36}$$

The recombination lifetime τ_r is a time constant defining the duration until the minority carrier density will return to thermal equilibrium.

We will now consider the opposite situation, where all charge carriers are removed from the semiconductor, for example by applying an external voltage. In this case, the initial recombination rate is zero and the generation rate will be equal to the thermal generation rate. With similar considerations as before, the time constant for the return to equilibrium is the generation lifetime τ_g:

$$\tau_g = \frac{n_i}{G_{th}} = \frac{1}{\beta n_i}, \tag{1.37}$$

which is different from the recombination lifetime.

For indirect semiconductors, such as silicon, the relationship between the two lifetimes is more complicated. This is due to the different crystal momentum for holes at the maximum of the valence band and electrons at the minimum of the conduction band, as already explained in the previous section. To enable the transfer of momentum to the crystal lattice, recombination in indirect semiconductors happens in a two step process, involving additional energy states in the forbidden band gap. These states are created by impurities and defects in the crystal lattice, which can capture and subsequently release an electron or hole depending on their type.

1. Basics on Silicon Semiconductor Technology

The process of capturing a hole or electron is called trapping and the defects in the silicon crystal enabling such processes are referred to as trapping centers. They cause a reduction of the signal induced by ionising radiation in a silicon sensor, as the trapped charges are captured and only emitted later. The readout electronics measures the signal carried by the non trapped carriers while the trapped charges are released too late. Radiation damage can increase the number of trapping centers and thus reduce the SNR of the sensor. See section 1.6 for more information.

1.5. Basic Semiconductor Structures

Using the n-type and p-type semiconductor materials described in the previous sections and adding materials like insulators and metals, we can create the main building blocks needed to form a silicon sensor. We will describe these basic structures in this section and discuss their most important properties.

1.5.1. The p-n Junction or Diode

The simplest semiconductor device is the diode. It can be described as an electronic check valve, enabling the flow of an electric current only in one direction. An equal description would be, that a diode has a low resistance if operated in conducting direction, while being high resistive if operated in reverse direction.

A p-n junction is created by joining a p-type and an n-type material. Beginning at the initial contact of two materials, the electrons from the n-type region will diffuse into the p-type region and recombine with the holes. This will create a space charge region near the junction, as the electrons diffusing from the n-type region will leave the donor ions uncompensated. Similarly the holes in the p-type region are compensated by the electrons and leave the acceptor ions uncompensated. The build-up of charge will counteract the diffusion until a certain depth of the space charge region is created, which is depleted of free charge carriers. The situation in the vicinity of the p-n junction is shown in figure 1.9.

The band structure in the p-type and n-type material is different prior to putting them into contact. Due to doping, the fermi level E_{fermi} will move towards the valence band for p-type material and towards the conduction band for n-type material. Upon contact of the two materials at thermal equilibrium, the fermi level at the p-n junction has to line up. This will shift the valence and conduction band and lead to a so called built-in voltage V_{bi} or diffusion voltage, as seen in figure 1.9 b. We can calculate V_{bi} using equations 1.12 and 1.13 and setting

1.5. Basic Semiconductor Structures

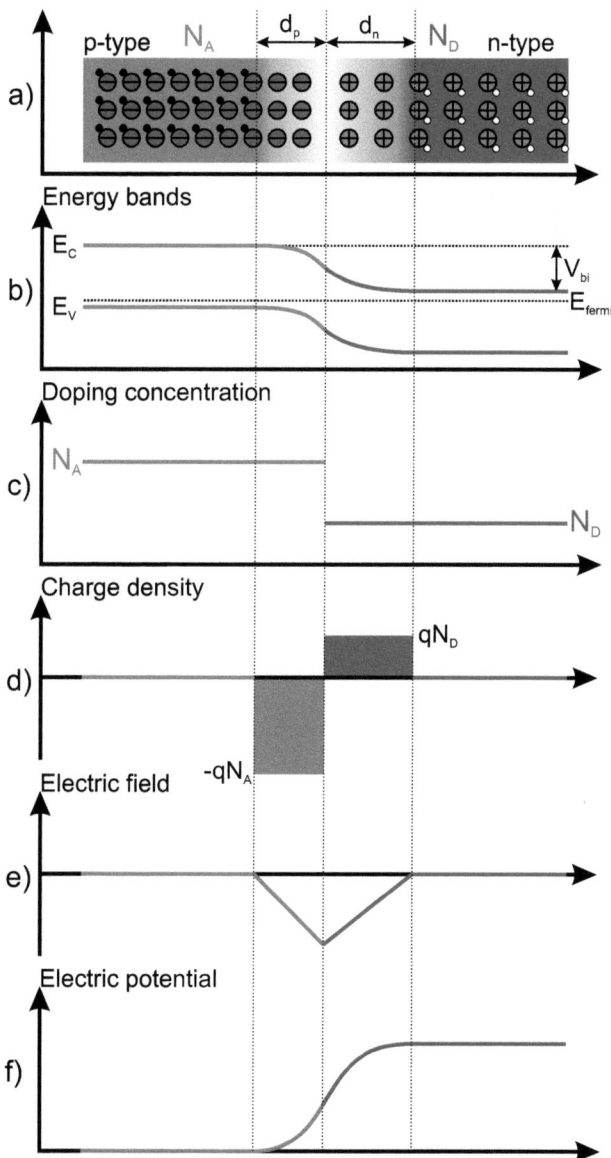

Figure 1.9.: Various parameters in the vicinity of a p-n junction. In diagram a) the donor and acceptor distribution of a partially depleted p-n junction is depicted. Diagram b) shows the energy band structure and c) the concentration of doping atoms. In diagram d) the charge density is shown, while e) and f) show the corresponding electric field and potential.

1. Basics on Silicon Semiconductor Technology

the majority carrier concentrations equal to the acceptor and donor concentrations:

$$n_n = N_D, \qquad p_p = N_A \qquad (1.38)$$

$$N_D \cdot N_A = n_i^2 \exp \frac{E_i^p - E_i^n}{kT} \qquad (1.39)$$

$$V_{bi} = \frac{1}{q}(E_i^p - E_i^n) = \frac{kT}{q} \ln \frac{N_A N_D}{n_i^2}. \qquad (1.40)$$

The size of the depleted zones in the p-type (d_p) and n-type (d_n) material can be calculated according to [1] as:

$$d_n = \sqrt{\frac{2\varepsilon\varepsilon_0}{q_e} \frac{N_A}{N_D(N_A + N_D)} V_{bi}} \qquad (1.41)$$

$$d_p = \sqrt{\frac{2\varepsilon\varepsilon_0}{q_e} \frac{N_D}{N_A(N_A + N_D)} V_{bi}} \qquad (1.42)$$

$$d = d_n + d_p = \sqrt{\frac{2\varepsilon\varepsilon_0(N_A + N_D)}{q_e N_A N_D} V_{bi}} \qquad (1.43)$$

For practical applications, p-n junctions are usually formed using a low doped material where a certain region gets highly doped towards the opposite type. The p-n junction is formed at the edge of the highly doped region which is in contact to the surrounding bulk with low doping concentration of opposite dopands. In this realistic case where the doping concentration on one side of the junction is significantly higher than on the opposite side (e.g. $N_A \gg N_D$), equation 1.43 becomes:

$$d = \sqrt{\frac{2\varepsilon\varepsilon_0}{q_e N_D} V_{bi}} \qquad (1.44)$$

If an external voltage is applied across the p-n junction, the charge carriers will start to drift according to the electric field. Depending on the polarity of the applied voltage, the width of the space charge region will shrink (*forward bias*) or expand (*reverse bias*) and we have to replace the built-in voltage V_{bi} with $V_{bi} - V$:

$$d = \sqrt{\frac{2\varepsilon\varepsilon_0(N_A + N_D)}{q_e N_A N_D}(V_{bi} - V)}. \qquad (1.45)$$

1.5. Basic Semiconductor Structures

For the realistic case as discussed above with $N_A \gg N_D$, equation 1.45 simplifies to:

$$d = \sqrt{\frac{2\varepsilon\varepsilon_0}{q_e N_D}(V_{bi} - V)}. \tag{1.46}$$

Current - Voltage Characteristics

In equilibrium, the diffusion of electrons (holes) due to the unequal majority charge carrier concentrations in the p-type and n-type materials is counterbalanced by the drift of electrons (holes) in the opposite direction induced by the space charge. An external voltage applied to the p-n junction will disturb this balance and influence the drift and diffusion currents of the charge carriers.

In forward bias direction, the lower electric field in the depleted region will reduce the drift of electrons (holes) from the p-side (n-side) to the n-side (p-side). In turn, the diffusion of electrons (holes) from n-side (p-side) to the p-side (n-side) is enhanced as minority carriers are injected into the junction. Under reverse bias, the electric field is increased and the diffusion of charge carriers is reduced resulting in a very small reverse bias current.

The total current density through the junction of an *ideal diode* can be described by the Shockley equation:

$$J = J_0 \left(\exp\frac{q_e V}{k_B T} - 1\right) \tag{1.47}$$

where the current in reverse bias direction ($V \ll 0$) is saturated at saturation current density $-J_0$ while it rises exponentially with the applied voltage in forward bias direction, as seen in figure 1.10. The saturation current density is given by [2]:

$$J_0 = \frac{q_e D_p p_{n0}}{L_p} + \frac{q_e D_n n_{p0}}{L_n} \tag{1.48}$$

where D_p and D_n are the diffusion constants for electrons and holes (see equations 1.25 and 1.24), p_{n0} and n_{p0} are the electron densities in the n-side and the hole densities in the p-side at thermal equilibrium and $L_p = \sqrt{D_p \tau_p}$ and $L_n = \sqrt{D_n \tau_n}$ are the diffusion lengths of holes and electrons.

Capacitance - Voltage Characteristics

We will now derive the capacitance - voltage characteristics for a one dimensional doping profile. As already discussed for equation 1.44, we will consider a simplified but realistic case for sensors, where a very small region of highly dope p-type material is embedded in a n-type

1. Basics on Silicon Semiconductor Technology

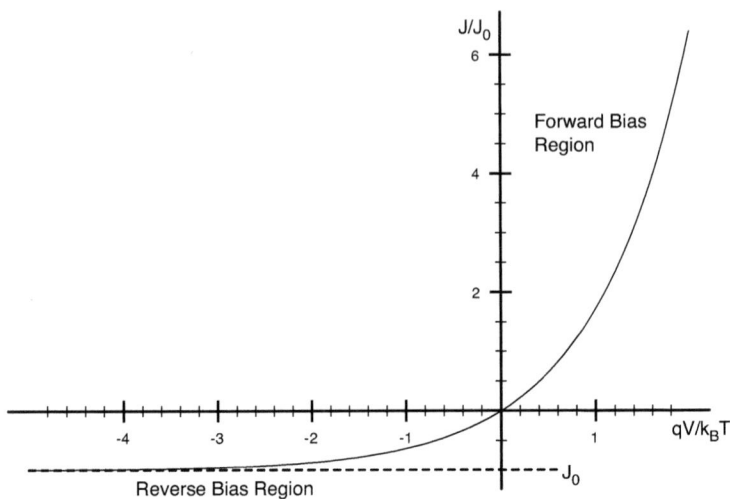

Figure 1.10.: Ideal current - voltage characteristics of a p-n junction.

substrate. In such an extremely asymmetric case, the space charge region is fully contained in the low doped n-type substrate.

To find the capacitance of the space-charge region we need the derivative of the charge Q_{SCR} with respect to the applied reverse bias voltage V:

$$C = \frac{dQ_{SCR}}{dV} = \frac{dQ_{SCR}/dW}{dV/dW} \tag{1.49}$$

Increasing the space-charge region by dx will increase the surface charge by $q_e N_D dx$, the electric field by $\frac{q_e N_D dx}{\varepsilon_0 \varepsilon}$ and the surface voltage by $\frac{q_e N_D dx}{\varepsilon_0 \varepsilon} x$. Integrating over the full space-charge region going from 0 to W we get the charge and voltage in the region:

$$Q_{SCR} = -\int_0^W q_e N_D(x) dx \tag{1.50}$$

$$V_{SCR} = \int_0^W \frac{q_e N_D(x)}{\varepsilon_0 \varepsilon} x \, dx \tag{1.51}$$

We can now calculate the derivatives of the charge and the voltage across the the space-charge region with respect to the size of the region. The voltage drop across the space-charge region is defined by the external voltage, reduced by the built-in voltage V_{bi} defined in equation

1.40. We now get:

$$\frac{dQ_{SCR}}{dW} = -q_e N_D(W) \tag{1.52}$$

$$\frac{dV}{dW} = \frac{dV_{bi}}{dW} - \frac{dV_{SCR}}{dW} = \frac{k_B T}{q} \frac{1}{N_D(W)} \frac{dN_D(W)}{dW} - \frac{q_e N_D(W)}{\varepsilon_0 \varepsilon} \tag{1.53}$$

For the inverse capacitance of the junction we get:

$$\frac{1}{C} = \frac{W}{\varepsilon_0 \varepsilon} - \frac{k_B T}{q} \frac{1}{q_e N_D^2} \frac{dN_D}{dW} \approx \frac{W}{\varepsilon_0 \varepsilon}. \tag{1.54}$$

As the variation of the built-in voltage with doping is negligible in comparison to the applied reverse bias voltage which is usually much higher, we can the corresponding term in equation 1.54.

The thickness of the space-charge region can now be derived from a capacitance measurement of the device as:

$$d = W = \frac{\varepsilon_0 \varepsilon}{C}. \tag{1.55}$$

Equation 1.55 is for a one dimensional p-n junction, while for a real diode with a junction area A the depletion depth becomes:

$$d = W = A\frac{\varepsilon_0 \varepsilon}{C}, \tag{1.56}$$

which resembles the situation of a parallel plate capacitor with electrodes of size A at a distance d and a dielectric in between with a dielectric constant of ε.

The doping concentration can be derived from the derivative of the inverse square of capacitance with the applied reverse bias voltage. Using equations 1.55 and 1.51 and again ignoring V_{bi} we get:

$$\frac{d(1/C^2)}{dV} = \frac{d(1/C^2)/dW}{dV/dW} = \frac{2W/(\varepsilon_0 \varepsilon)^2}{q_e N_D W/(\varepsilon_0 \varepsilon)} = \frac{2}{q_e N_D \varepsilon_0 \varepsilon}, \tag{1.57}$$

and the doping concentration at depth W is:

$$N_D = \frac{2}{q_e \varepsilon_0 \varepsilon \frac{d(1/C^2)}{dV}} \tag{1.58}$$

Full Depletion Voltage

The voltage at which the depleted space-charge region covers the full depth of the device is called *Full Depletion Voltage* V_{FD}. Following equation 1.58 the capacitance of the device will

1. Basics on Silicon Semiconductor Technology

increase with the reverse bias voltage V until $V = V_{FD}$. For $V \geq V_{FD}$ it will remain constant as seen in figure 1.11.

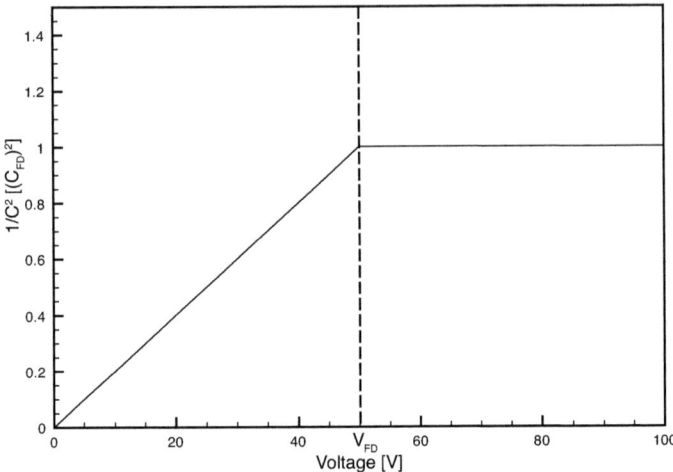

Figure 1.11.: Plot of a C-V curve of an ideal diode. Before full depletion, the inverse square capacitance rises linearly. At $V = V_{FD}$ the device is fully depleted and the capacitance remains constant.

Following equation 1.46 the thickness d of a device can be calculated from V_{FD} for $V \gg V_{bi}$:

$$d = \sqrt{\frac{2\varepsilon_0 \varepsilon}{q_e N_D} V_{FD}} \qquad (1.59)$$

For practical reasons the doping concentration N_D is substituted with the resistivity of the bulk material from equation 1.15 which yields:

$$d = \sqrt{2\varepsilon_0 \varepsilon \mu_n \rho V_{FD}} \qquad (1.60)$$

In comparison to an ideal diode, the p-n junctions in strip sensors are segmented, which makes the electric field inside the bulk non-linear due to edge effects. The solutions to the Poisson equation describing the voltage drop are more complicated and are discussed in detail in [4]. The solution for strip sensors is given by:

$$V_{FD,sensor} = V_{FD,diode} \left[1 + 2\frac{p}{d} f\left(\frac{w}{p}\right) \right], \qquad (1.61)$$

1.5. Basic Semiconductor Structures

where p is the pitch between neighbouring strips and w is the width of the strip. The function $f(w/p)$ is a numerical approximation for the solution and is according to [4]:

$$f(x) = -0.00111x^{-2} + 0.0586x^{-1} + 0.240 - 0.651x + 0.355x^2 \quad (1.62)$$

For a sensor with a standard width-to-pitch ration of 0.25, the correlations between resistivity, thickness and full depletion voltage are:

$$d = \sqrt{\frac{2\varepsilon_0\varepsilon\mu_n\rho V_{FD,sensor}}{1+2\frac{p}{d}0.3161}} \quad (1.63)$$

$$\rho V_{FD,diode} = \frac{\rho V_{FD,sensor}}{1+2\frac{p}{d}0.3161} = \frac{d^2}{2\varepsilon_0\varepsilon\mu_n} \quad (1.64)$$

$$\rho V_{FD,sensor} = \frac{d^2(1+0.6322\frac{p}{d})}{2\varepsilon_0\varepsilon\mu_n} = \frac{d^2+0.6322pd}{2\varepsilon_0\varepsilon\mu_n} \quad (1.65)$$

1.5.2. The n+-n or p+-p Junction

In the previous chapter on p-n junctions, the build-up of a space-charge region was explained due to electrons diffusing from the n into the p region while holes are diffusing from p to n. The diffusion of electrons and holes stops, when the electric field inside the space-charge region is high enough. The electrical potential of the space-charge region is called built-in voltage V_{bi}.

A similar situation is created in single-type semiconductors if the doping concentration changes significantly over a small area[1]. In the case for n+-n junctions, electrons from the heavily doped n+ region diffuse into the much less doped n region and create a space charge region. The resulting electrical field counteracts the diffusion and the Fermi levels of the two regions in the band model are lined up. This again results in a built-in voltage V_{bi}.

1.5.3. The Metal - Semiconductor Contact

Historically, the metal-semiconductor contact was one of the first semiconductor devices showing rectifying properties. As the name suggests, such contacts form at the junction between a metal and a semiconductor of low doping concentration. Such contacts are also called Schottky barriers.

The situation at the metal - semiconductor junction is similar to a p-n junction. The Fermi level E_{fermi} in the metal and the semiconductor are at different energies. Upon contact, these

---

[1] Heavily doped regions are denoted as n+ or p+, hence the name n+-n or p+-p for single-type junctions.

the levels have to line up and therefore alter the band structure creating barriers for the charge carriers.

Let's consider the situation of separated metal and n-type semiconductor. The energy necessary to move an electron from the Fermi level into the vacuum is described by the work function $q_e\Phi_m$. The specific value of Φ_m depends on the type of metal and is different to the value $\Phi_s < \Phi_m$ in the semiconductor which in turn depends on its type and doping. Furthermore we define the electron affinity $q_e\chi$ which is the difference between the conduction bands edge to the vacuum level in the semiconductor. This value does not depend on the doping of the material.

Now we consider the metal and the semiconductor in direct contact, where the Fermi levels in both materials have to line up and a built-in voltage $V_{bi} = \Phi_m - \Phi_s$ is created, as the bands will be bent in the vicinity of the contact. Charge carriers in the semiconductor will move to fulfill the thermal equilibrium condition at any point, while the situation in the metal is somewhat different. The number of free electrons in the metal is so huge, that the electric field inside the metal is approximately zero. Therefore the positive space charge region created in the semiconductor will be compensated by a surface charge at the metal side. This creates a barrier for electrons diffusing from the metal to the semiconductor side. The height of the barrier can be described by:

$$q_e\Phi_{Bn} = q_e(\Phi_m - \chi) \tag{1.66}$$

This barrier is not influenced by an external bias voltage, while the threshold for electrons traveling in the opposite direction will change. This rectifying function is similar to a diode with a low forward voltage drop. For strip sensors, Schottky barriers are created at the interface of the semiconductor material to the metallisation for external connections. Their rectifying properties are a nuisance and should be avoided by using highly doped silicon at the interface. The high doping concentration will decrease the width of the potential barrier and tunneling of electrons becomes more important as described in [2]. The characteristic resistance in the metal to semiconductor direction will be negligible and the contact will show an ohmic behaviour.

1.5.4. The Metal - Oxide - Semiconductor Structure

The **M**etal - **O**xide - **S**emiconductor (MOS) structure is a very important component widely used in the semiconductor industry. One example utilizing MOS structures are **C**harge **C**oupled **D**evices (**CCD**) which are used as imaging sensors. For silicon strip sensors, MOS structures are usually only used as test structures to provide access to the measurement of interface properties.

1.5. Basic Semiconductor Structures

A MOS structure consists of a semiconductor covered by an insulator which is often made of silicon dioxide (SiO$_2$) but can be composed of any material with insulating properties. To understand the functionality of the structure, we have to understand the distribution of charge carriers inside the semiconductor device and at the interface between semiconductor and insulator for different external voltages applied between the metal contact and the semiconductor. We will first consider a simplified situation for a n-type semiconductor in thermal equilibrium, neglecting any charge carriers in the oxide. Depending on the externally applied voltage we can distinguish four different situations: *flat band*, *accumulation*, *depletion* and *inversion*.

Flat Band Situation ($V = V_{FB}$)

Let us consider a more simplified situation first, where the work functions (the energy necessary to move an electron from the Fermi to the vacuum level) for metal and semiconductor are equal. Without the application of an external voltage, the electrons in the n-type semiconductor are uniformly distributed and the electric field will be zero everywhere along the device. No external voltage is needed to achieve this situation and the flat-band voltage V_{FB} is zero.

In a more realistic situation, the work functions for an electron in the metal and in the semiconductor are different, while the vacuum level is constant throughout the device, as we considered the oxide to be free of charges and therefore without an electric field. To achieve this situation we have to apply an external voltage V_{FB} between the metal electrode and the semiconductor to compensate the different work functions:

$$V_{FB} = \Phi_m - \Phi_s \tag{1.67}$$

The situation is represented in figure 1.12.

Accumulation Situation ($V > V_{FB}$)

If the externally applied voltage exceeds the flat band voltage, the bands at the oxide - semiconductor interface will bend down. Electrons will accumulate in a thin layer at the surface as the Fermi level moves closer to the conduction band edge. The thermal equilibrium condition must be fulfilled within the semiconductor:

$$\frac{n}{n_i} = \exp\frac{E_{fermi} - E_i}{k_B T} \tag{1.68}$$

1. Basics on Silicon Semiconductor Technology

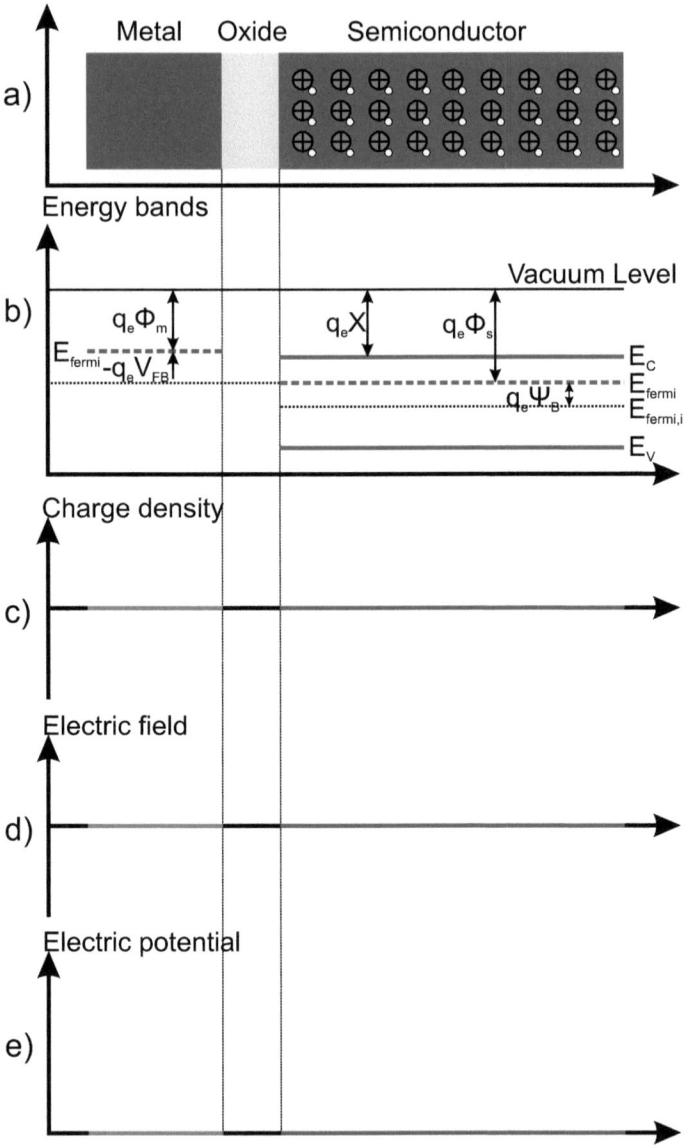

Figure 1.12.: An n-type MOS structure in the flat band situation $V = V_{FB}$. The oxide is considered to be free of charges.

1.5. Basic Semiconductor Structures

where n is the actual and n_i the intrinsic charge carrier concentration. The surface charge density is given by:

$$Q_{acc} = -\varepsilon_0 \varepsilon_{ox} \frac{V - V_{FB}}{d_{ox}} = -C_{ox}(V - V_{FB}) \qquad (1.69)$$

where C_{ox} is the capacitance per unit area of the oxide layer.
The situation is represented in figure 1.13.

Depletion Situation ($V < V_{FB}$)

If the externally applied voltage is smaller than the flat band voltage V_{FB}, the bands at the oxide-semiconductor interface will bend upwards. The concentration of electrons near the interface will decrease and a depleted region will build up.

According to [1], the depth of the depletion layer is given by:

$$d_s = \sqrt{\frac{\varepsilon_0 \varepsilon_s}{q_e N_D}(V_{FB} - V) + \left(\frac{\varepsilon_s}{\varepsilon_{ox}} d_{ox}\right)^2} - \frac{\varepsilon_s}{\varepsilon_{ox}} d_{ox} \qquad (1.70)$$

The situation is represented in figure 1.14.

Inversion Situation ($V \ll V_{FB}$)

If the external voltage is much smaller than the flat band voltage V_{FB}, such that the intrinsic level at the oxide-semiconductor interface $q_e \Phi_B$ is higher than the Fermi level $q_e \Phi_s$, a majority of holes will accumulate at the interface. For $\Phi_s = -2\Phi_B$ the hole density at the interface equals the electron density in the bulk of the semiconductor. This situation as also called *strong inversion*. The depletion depth reaches its maximum and even a further increase of the external voltage will not change the depth:

$$d_{max} = \sqrt{\frac{4\varepsilon_0 \varepsilon_s \Phi_B}{q_e N_D}}. \qquad (1.71)$$

The situation is represented in figure 1.15.

Oxide Charges

In a real MOS structure, composed of silicon and silicon dioxide, charges will be present inside the oxide and at the Si-SiO^2 interface which will change the flat band voltage V_{FB}. The change depends on the amount and distribution of the charges $\rho(x)$ inside the oxide and can be

1. Basics on Silicon Semiconductor Technology

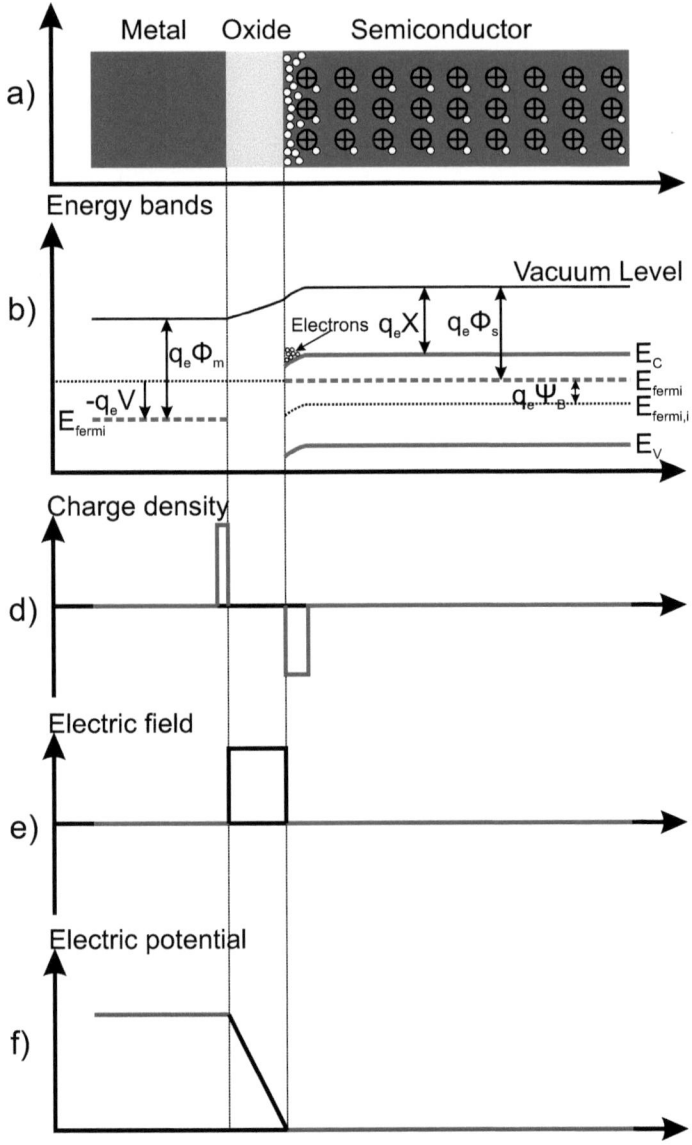

Figure 1.13.: An n-type MOS structure in the accumulation situation $V > V_{FB}$. The oxide is considered to be free of charges.

1.5. Basic Semiconductor Structures

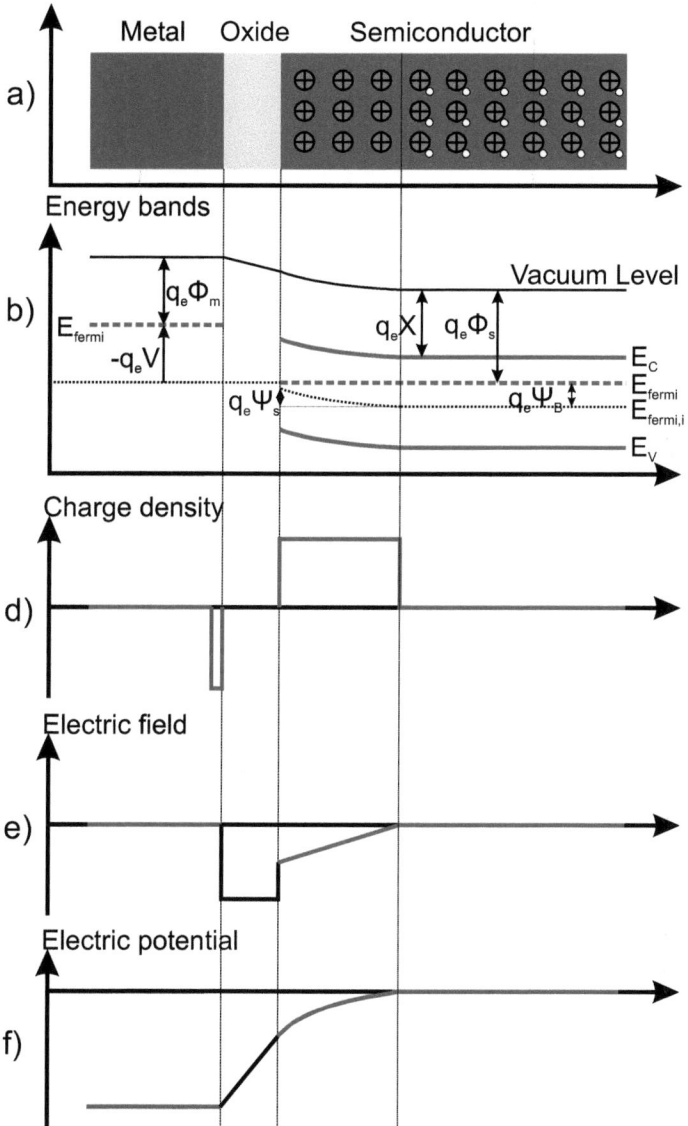

Figure 1.14.: An n-type MOS structure in the depletion situation $V < V_{FB}$. The oxide is considered to be free of charges.

1. Basics on Silicon Semiconductor Technology

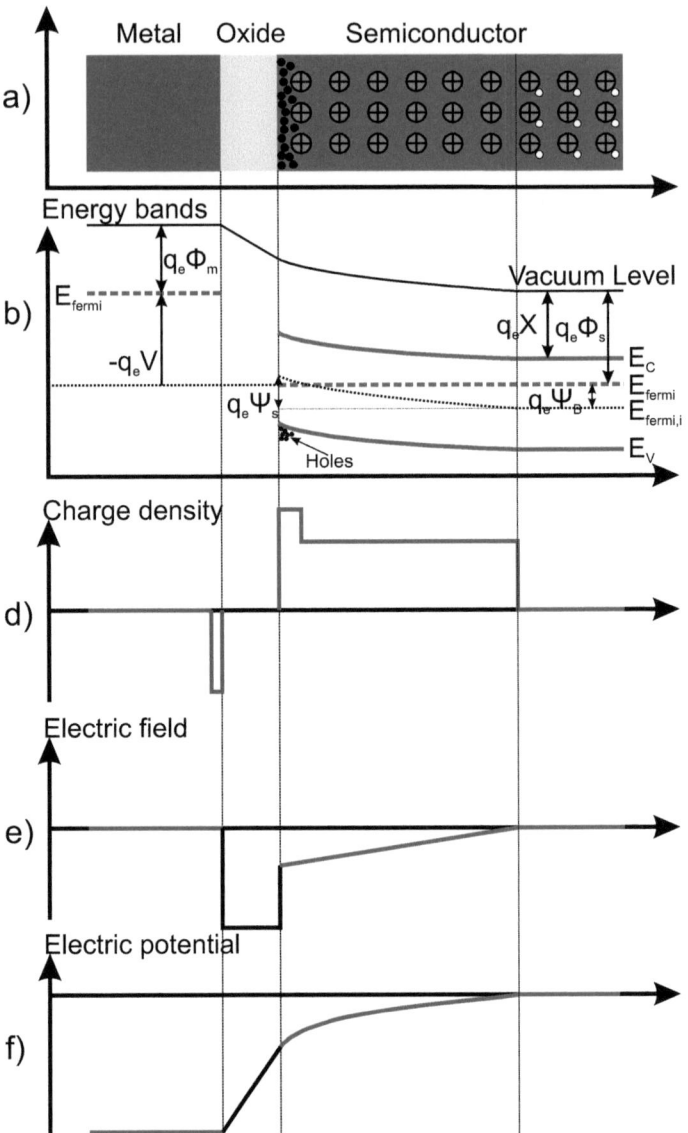

Figure 1.15.: An n-type MOS structure in the (weak) inversion situation $V \ll V_{FB}$. The oxide is considered to be free of charges.

1.5. Basic Semiconductor Structures

described as:

$$\Delta V_{FB,oxB} = \frac{1}{\varepsilon_0 \varepsilon_{ox}} \int_0^{d_{ox}} \rho(x) x \, dx. \quad (1.72)$$

The change due the interface region can be described as a thin layer of charges with a density of σ at a distance d_{ox} from the metal:

$$\Delta V_{FB,oxI} = -\frac{\sigma d_{ox}}{\varepsilon_0 \varepsilon_{ox}}. \quad (1.73)$$

Equation 1.67 has to be corrected for these to effects and becomes:

$$V_{FB} = \Phi_m - \Phi_s - \frac{1}{\varepsilon_0 \varepsilon_{ox}} \left[\sigma d_{ox} + \int_0^{d_{ox}} \rho(x) x \, dx \right]. \quad (1.74)$$

Capacitance - Voltage Characteristics

The capacitance of the MOS structure is measured by applying a small AC-voltage on top of the DC bias voltage. The response from the structure depends on the relationship between the frequency of AC-voltage and the time it takes for the structure to return to equilibrium. The measured capacitances are strongly frequency dependent.

For low frequencies we can assume ($C_{ox} = \frac{\varepsilon_0 \varepsilon_{ox}}{d_{ox}}$, $C_s = \frac{\varepsilon_0 \varepsilon_s}{d_s}$):

Accumulation region $C = C_{ox}$

Depletion region $C = \frac{C_{ox} C_s}{C_{ox} + C_s}$ as the capacitance of the oxide and the depletion layer appear in series

Inversion region $C = C_{ox}$ as the depletion layer stays constant, while the surface charge density varies with the applied voltage.

At high frequencies the measured capacitance again appears as $C = \frac{C_{ox} C_s}{C_{ox} + C_s}$, because the depletion layer depth varies with the voltage while the surface charge density stays constant.

1.5.5. The Polysilicon Resistor

Resistors of a wide range of resistances can be created by depositing and doping a layer of polysilicon. Pure polysilicon has a resistivity of 230 kΩcm which can be adjusted to the appropriate value by doping. Special attention has to be made when contacting polysilicon resistors with metal due to the Schottky barrier described in section 1.5.3. Usually a good electrical contact between metal and the resistor is achieved using a heavily doped area of polysilicon to make the contact to the metal. This structure is also called a *Polysilicon Head*.

1.6. Radiation Damage and the NIEL hypothesis

The effects of radiation on silicon sensor have to be carefully evaluated to ensure proper operation over the full length of an experiment's expected lifetime. Even today, most effects are only partly understood. We have to rely on careful evaluation of radiation experiments and parametrization of the resulting effects, without a complete understanding about the physical background.

The silicon sensor have to endure a variety of radiation types, which, for simplicity, we are going to divide into two types:

- Charged particles like protons, pions, electrons, ...
- Neutral Particles like neutrons

Neutral particles interact by elastic or inelastic scattering with the semiconductor nucleus, while charged particles like protons or electrons scatter by electrostatic interaction as well. The mass of the scattering particle is also important, because it limits the maximum energy that can be transferred from the incident particle to a lattice atom. Electrons, for example, seldom create lattice defects as they cannot transfer enough energy and cause point defects and ionization only. In table 1.2 some important characteristics of the primary interactions of radiation are given. The concepts of point and cluster defects will be clarified later.

Radiation Interaction	Electrons	Protons	Neutrons	Si^+
	Coulomb scattering	Coulomb and nuclear scattering	Elastic nuclear scattering	Coulomb scattering
$T_{max}[eV]$	155	133,700	133,900	1,000,000
$T_{av}[eV]$	46	210	50,000	265
$E_{min}[eV]$				
point defect		260,00	190	190	25
defect cluster	4,600,000	15,000	15,000	2,000

Table 1.2.: Characteristics of interaction of radiation with silicon [1]. The radiation energy is 1 MeV, T_{max} is the maximum kinematically possible recoil energy, T_{av} the mean recoil energy and E_{min} the minimum radiation energy needed for the creation of a point defect and for a defect cluster.

Figure 1.16 shows the <u>N</u>on <u>I</u>onizing <u>E</u>nergy <u>L</u>oss (**NIEL**) for different particle energies. The damage caused by different types of radiation can be compared using the *hardness factor* κ which is defined according to [5]:

$$\Phi_{eq}^{1MeV} = \kappa \Phi \tag{1.75}$$

1.6. Radiation Damage and the NIEL hypothesis

where κ is defined as:

$$\kappa = \frac{EDK}{EDK(1MeV)} \tag{1.76}$$

with EDK the Energy spectrum averaged Displacement KERMA[2]

$$EDK = \frac{\int D(E)\Phi(E)dE}{\int \Phi(E)dE} \tag{1.77}$$

where $\Phi(E)$ is the differential flux and

$$D(E) = \sum_k \sigma_k(E) \int dE_R f_k(E, E_R) P(E_R) \tag{1.78}$$

is the displacement KERMA or the damage function for the energy E of the incident particle, σ_k the cross section for reaction k, $f_k(E, E_R)$ the probability of the incident particle to produce a recoil of energy E_R in reaction k and $P(E_R)$ the partition function (the part of the recoil energy deposited in displacements). The normalised EDK for 1 MeV neutrons is $EDK(1\ \text{MeV})=95$ MeVmb. The integration is done over the full energy range.

The hardness factor κ now becomes:

$$\kappa_{1\ MeV\ n} = \frac{\int D(E)\Phi(E)dE}{95\ MeVmb \cdot \Phi} \tag{1.79}$$

Using the hardness factor $\kappa_{1\ MeV\ n}$, the NIEL hypothesis allows the scaling of the damage caused by any charged particle to 1 MeV neutrons.

1.6.1. Bulk and Surface Damage

In this section we are going to investigate the effects of the two types of radiation (neutral and charged particles) on the semiconductor lattice and the insulating oxide layer. As we have learned from the preceding section, the interaction depends on whether the particles are charged or not and on their mass.

The actual detection process occurs in the depleted space-charge region of the silicon bulk. Imperfections in the crystal lattice would influence the detector properties, generally degrading the sensor performance. Defects created by radiation which dislocate silicon atoms from their lattice site are the main concern. As mentioned before, heavy particles scatter with the silicon nucleus and transfer kinetic energy to it. If the energy transferred exceeds about 15 eV, a

---

[2]Kinetic Energie Released per unit MAss (**KERMA**) is the sum of the initial kinetic energies of all the charged particles liberated by uncharged ionizing radiation (neutrons and photons) in a sample of matter, divided by the mass of the sample. It is measured in Gray.

1. Basics on Silicon Semiconductor Technology

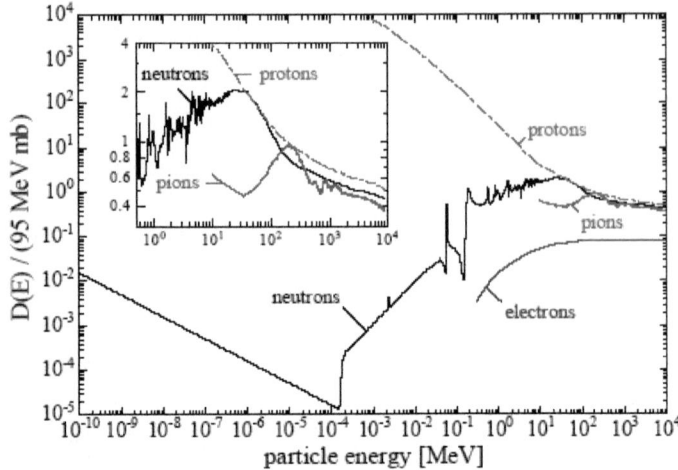

Figure 1.16.: Non ionizing energy loss of different particles [6]. The cross sections are normalized to 1 MeV neutrons of equivalent fluence Φ_{eq}.

dislocation of a lattice atom is possible. More specifically, at a recoil energy of about 25 eV, the probability of displacement of a silicon lattice atom is roughly one half.

Recoil energies below 1 - 2 keV create only isolated point defects, between 2 keV and 12 keV the energy is high enough to create one defect cluster and additional point defects and above 12 keV several clusters and additional point defects can be produced. A cluster is a dense agglomeration of point defects that appear at the end of a recoil track where the incident particle loses its last 5 - 10 keV of kinetic energy and the elastic scattering cross-section increases by several orders of magnitude.

In the insulating oxide layer which separates the p-type silicon from the aluminium readout strips, the situation is different. The structure is already highly irregular, therefore the interaction of radiation with the nucleus can be ignored. The additional damage to the oxide structure will not alter the properties of that region.

Much more important is the ionization caused by charged particles, and by photons. One may consider the oxide as a region with a high density of defects whose charge state can be altered by irradiation. New electrons and holes are created in the oxide layer. The electron mobility is several orders of magnitude larger than that of holes. Compared to holes, radiation-generated electrons will diffuse out of the insulator in relatively short time and the capture of holes is the dominant process that changes the oxide's properties. Radiation damage of oxide manifests itself as a buildup of positive charge due to semipermanent trapping centers, which causes a shift in the flat-band voltage which can be measured.

1.6.2. Changes in Properties due to Defect Complexes

The defects in detector bulk material are still mobile at room temperature. Part of those defects will even vanish either by an interstitial filling a vacancy or by diffusing out of the surface. However, they may become stable by interacting with another radiation induced defect, or with an imperfection from the crystal growth process. Let's have a look on the effects and changes in detector properties that are caused by these stable defects. Before going into depth, it is worth mentioning that these defects can have one of the following main consequences:

- Alterations of charge density in the space - charge region
- Formation of recombination - generation centers
- Formation of trapping centers

Alterations of Charge Density in the Space - Charge Region

The original dopants such as phosphorus or boron may be captured into new defect complexes, thereby loosing their original function as flat donors or acceptors. They may assume a charge state different from the original one. This will change the effective doping of the semiconductor. In figure 1.17 the fluence dependence of the effective doping of an originally n-type silicon can be seen. The dashed lines correspond to the parametrisation of that effect. At a certain fluence, the silicon even changes it's doping state from n-type to effectively p-type. This effect is called *type inversion*.

The effective doping concentration can be parametrised by:

$$N_{eff}(\Phi) = N_{D,0} e^{a\Phi} - N_{A,0} - b\Phi \tag{1.80}$$

with $N_{D,0}$ and $N_{A,0}$ being the donator and acceptor concentration before irradiation and a and b constants to be determined experimentally.

A consequence of the alterations in the space charge region is the change of the full depletion voltage V_{FD} and therefore the operating voltage of the sensor.

Recombination - Generation Centers These defects can capture and emit electrons or holes and raise the volume-generated leakage current. The leakage current as a function of the

1. Basics on Silicon Semiconductor Technology

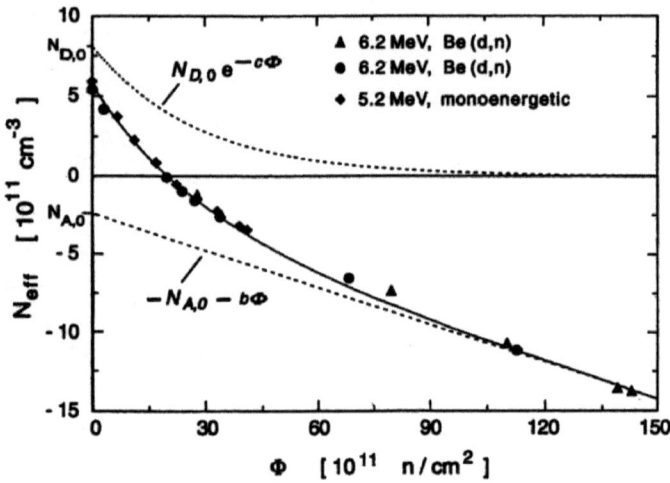

Figure 1.17.: Fluence dependence and parameterisation of the effective doping according to equation 1.80 for an n-type silicon wafer irradiated with neutrons [7]. The data has been corrected for self-annealing during the extended irradiation period.

fluence is shown in figure 1.18. A linear relationship between current and fluence is found. It can be parameterized as:

$$\frac{\Delta I_{vol}}{V} = \alpha \Phi, \qquad (1.81)$$

where the leakage current damage parameter α is according to [8]:

$$\alpha = (3.99 \pm 0.03) \times 10^{-17} A/cm. \qquad (1.82)$$

Trapping Centers Another problem is the trapping of an electron or hole and remission a short time afterwards. The charge is released too late to contribute to the signal. The signal measured by the readout electronics is smaller while exhibiting more noise resulting in a reduced SNR.

We present a global parameterization of the fluence dependence of the trapping time:

$$\frac{1}{\tau_t} = \frac{1}{\tau_{t0}} + \gamma \Phi \qquad (1.83)$$

where τ_t is the average time a hole/electron stays trapped and τ_{t0} is the value before radiation.

1.6. Radiation Damage and the NIEL hypothesis

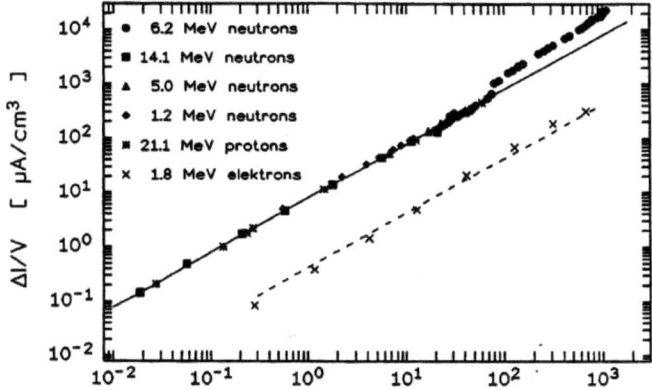

Figure 1.18.: A 1 MeV neutron equivalent fluence dependence and parameterization of the volume-generated current for an n-type silicon wafer irradiated with neutrons [7]. The data has been corrected for self-annealing occurring during the extended irradiation period.

This parameterization works well for hole and electron trapping at moderate fluence. A value of $\gamma \approx 0.24 \times 10^{-6}$ cm²/s is found

For electrons at high fluences, a steeper increase of trapping probability was observed (see figure 1.19).

1.6.3. Annealing

Observing a sensor after an extended period of exposure to irradiation, one notices that the damage effects diminish with time. This effect is called *annealing*. The rate of damage decrease is strongly dependent on temperature.

This effect can naively be interpreted as diffusion of radiation induced crystal defects out of the detector bulk and, more importantly, by the recombination of vacancies and interstitials. Still, one must keep in mind that annealing is a rather complicated process involving many different and only partially understood interactions between defects and defect complexes.

In figure 1.20 the annealing of the radiation induced changes of effective doping is shown over an extended period of time at constant room temperature (20°C). One may notice an inversion of the annealing effect on the time scale of months.

43

1. Basics on Silicon Semiconductor Technology

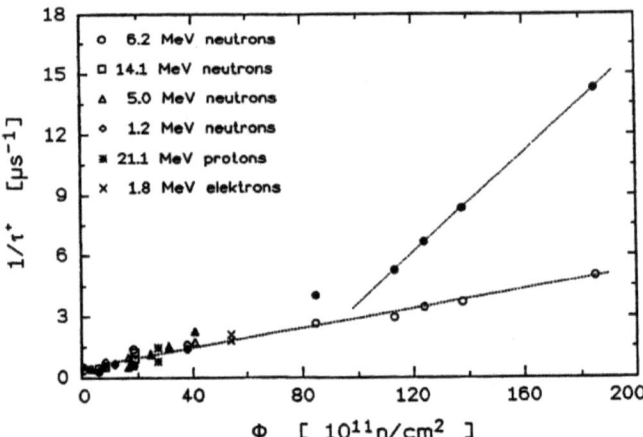

Figure 1.19.: A 1MeV neutron equivalent fluence dependence and parameterization of the inverse trapping time constant for holes (open symbols) and electrons (solid symbols) for an n-type silicon wafer irradiated with neutrons [7].

1.6.4. Reverse Annealing

The inversion of the annealing process is called *reverse annealing*. Following the initial annealing of irradiated silicon sensors, the effective doping concentration is increasing again after a few weeks to months at room temperature. This surprising effect can be explained as the transformation of radiation induced electrically inactive defect complexes into electrically active ones. Reverse annealing is also strongly temperature dependent. Below approximately 0°C the process is almost completely suppressed.

1.6. Radiation Damage and the NIEL hypothesis

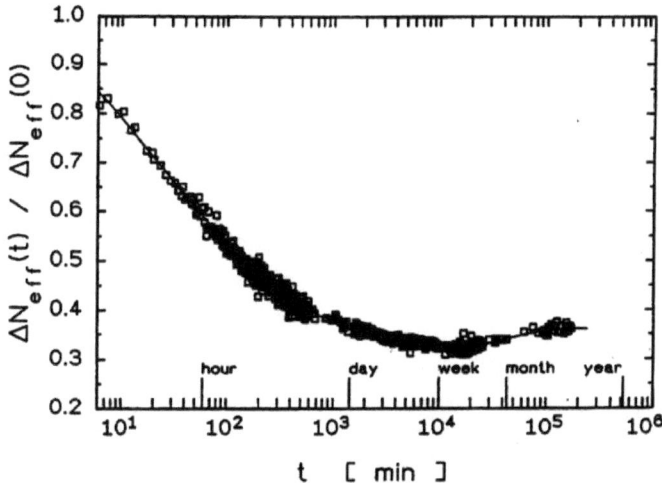

Figure 1.20.: Room temperature annealing of radiation induced changes of effective doping [7]. Data are corrected to correspond to a short irradiation followed by a longterm observation at constant (20°C) temperature. A decrease of the effective doping concentration can be observed on the time scale of weeks (annealing), followed by a rise (reverse annealing) on the time scale of months after irradiation.

A semiconductor detector is a device that uses a semiconductor (usually silicon or germanium) to detect traversing charged particles or the absorption of photons. In the field of particle physics, these detectors are usually known as silicon detectors. When their sensitive structures are based on a single diode, they are called semiconductor diode detectors. When they contain many diodes with different functions, the more general term semiconductor detector is used. Semiconductor detectors have found broad application during recent decades, in particular for gamma and X-ray spectrometry and as particle detectors.

Wikipedia on **Semiconductor detector**

2
Silicon Strip Sensors

Silicon strip sensors are the favored tracking devices in most modern high energy physics experiments. The basic design has not changed much over the last few centuries since the first strip sensor in planar technology was developed for the NA11 experiment at CERN[9]. Nevertheless, improvements were made to the designs and the production techniques used

2. Silicon Strip Sensors

to manufacture such devices. The current status of commonly used design and production principles will be reviewed in this chapter.

2.1. Working Principle

The main purpose of a strip sensor is the precise spatial tracking of ionizing particles. The effect used to create a measurable signal from a particle passing the sensor is based on the electromagnetic interaction of the particle with the the silicon bulk. The resulting generation of a current inside the bulk was described in section 1.4.3 for charged particles and in section 1.4.2 for photons.

The amount of charge created by a single particle is very small compared to the intrinsic charge carrier density. With a mean energy loss of a MIP in silicon of $(dE/dx)_{mean} = 388 \; eV/\mu m$ ([10]) and a mean energy for electron-hole creation in silicon of $E_{pair} = 3.63$ eV (see table 1.1) the number of electron-hole pairs created by a MIP in $t_{Sensor} = 300$ µm silicon becomes:

$$\frac{(dE/dx)_{mean} \times t_{Sensor}}{E_{pair}} = \frac{388 \times 300}{3.63} \approx 32 \times 10^3 \; e^-h^+ \, pairs. \tag{2.1}$$

The intrinsic charge carrier generation at $T = 300$ K for silicon is $n_i = 1.45 \times 10^{10}$. For silicon with the same thickness of $t_{Sensor} = 300$ µm as before and an area of $A_{Sensor} = 1$ cm^2 this amounts to:

$$n_i \cdot t_{Sensor} \cdot A_{Sensor} = 1.45 \times 10^{10} \cdot 300 \cdot 1 \times 10^{-4} \approx 4.4 \times 10^8. \tag{2.2}$$

The signal created by a MIP would be four orders of magnitude lower than the intrinsic noise created by 1 cm^2 of silicon. Therefore it is necessary to significantly reduce the number of intrinsic charge carriers. One way would be to operate the sensor at very low temperatures, but a much more practical way is the usage of a p-n junction operated in reverse bias mode.

The main sensor element of modern strip sensor is the depleted zone of a p-n junction. Figure 2.1 shows the basic working principle where the p-n junction is operated in reverse bias mode. For practical reasons, the junction is very asymmetric while making the depleted region occupy most or the full depth of the silicon substrate. Therefore the silicon bulk is made of pre-doped material (usually n-type) and the junction is formed by heavy doping of a small surface layer (usually p+-type). To make the full depth of the bulk sensitive, it is necessary to fully deplete it by a sufficiently high reverse bias voltage.

When an ionising particle particle traverses the depleted region, it creates electron-hole pairs.

2.1. Working Principle

Due to the electric field inside the bulk, these charge carriers start to drift towards the electrodes (see section 1.3.1) and thereby induce an electric current which can be measured.

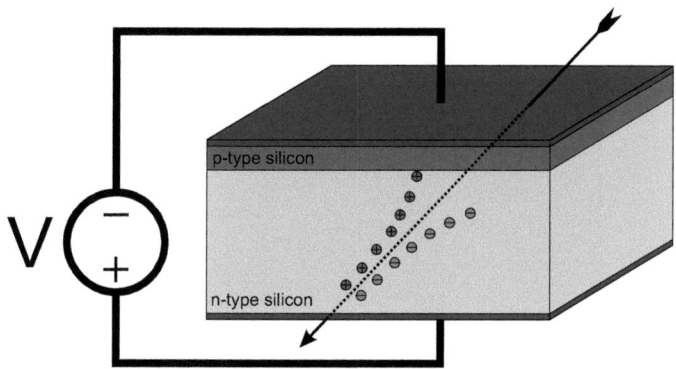

Figure 2.1.: Charge generation of an ionizing particle inside the depleted silicon bulk. The charges will drift to the electrodes due to the electric field and induce a current which can be measured.

The energy deposition in thin layers is considerably non-deterministic and the statistical fluctuations are very asymmetric following a Landau distribution as seen in figure 2.2. It was first described in [11] with a more elaborate discussion and corrections to the distribution in [12].

As calculated above, the mean number of electron-hole pairs is about 32.000 but due to the Landau like behaviour of the energy deposition, the most probable number of electron-hole pairs becomes with $(dE/dx)_{MPV} = 276 \ eV/\mu m$:

$$\frac{(dE/dx)_{MPV} \times t_{Sensor}}{E_{pair}} = \frac{276 \times 300}{3.63} \approx 23 \times 10^3 \ e^-h^+ pairs. \tag{2.3}$$

The most probable value is the number often used in literature.

The electron-hole pairs created by the incident particle drift towards the electrodes and create an electric current. The signal is induced by the movement of the charges through electromagnetic interaction and is proportional to the carrier mobilities according to Ramo's theorem [13]:

$$J_0 = \frac{q_e}{d} \left(\sum v_n + \sum v_p \right), \tag{2.4}$$

where d is the detector thickness, q_e the elementary charge and v_n and v_p the drift velocities for electrons and holes.

2. Silicon Strip Sensors

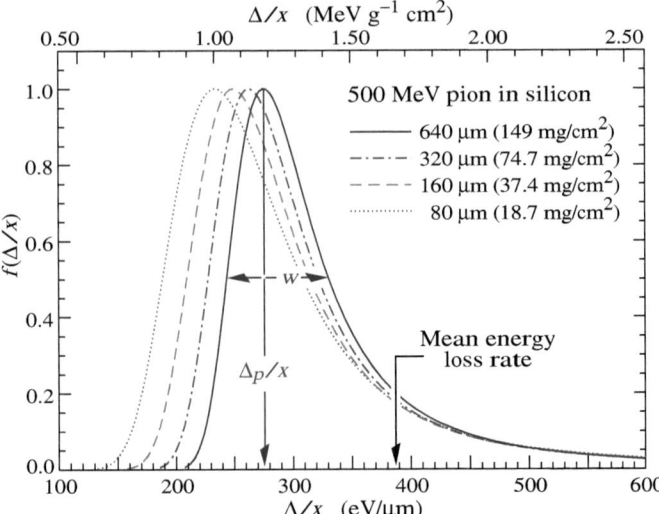

Figure 2.2.: Straggling functions in silicon for 500 MeV pions, normalised to unity at the most probable value δ_p/x [10]. The width w is the full width at half maximum.

The maximum charge which can be collected by integrating amplifiers, can be calculated by the time integral over the induced current:

$$Q_0 = \int_0^{t_{int}} J_0 dt = \int_0^{t_{int}} \frac{q_e}{d} \left(\sum v_n + \sum v_p \right) dt, \tag{2.5}$$

where t_{int} is the integration time of the amplifier. The measured signal can be reduced due to the trapping of charges inside the bulk. The Charge Collection Efficiency (**CCE**) of a sensor is an important parameter and a subject of interest when analysing the radiation hardness of a sensor:

$$CCE = \frac{Q_c}{Q_0}, \tag{2.6}$$

where Q_c is the charge collected in an actual measurement.

2.2. Design Basics of a Silicon Strip Sensor

Using the basic working principle described in the previous section, it is possible to detect a particle hitting the sensor. It does not give any additional information on the location of the hit on the sensor. This can be achieved by creating smaller sensing elements based on the same principle which are electrically isolated from each other but located on the same sensor. For practical reasons, we can distinguish two different approaches:

Pixel Sensor have very small quadratic or rectangular shaped sensor elements called pixels. Due to their almost symmetric layout, they provide two dimensional spatial resolution. Each of the pixel is readout by a dedicated amplifier which is in some designs also responsible for biasing it. Pixel sizes are usually very small, in the order of one hundred microns, which results in a very high number of pixels per sensor area. Due to the high number of readout channels, the costs per area covered by sensors is very high. Due to their inherent radiation hardness, they are used as the innermost tracking layers as in the CMS tracker (see section 3.1.1).

Strip Sensor have narrow and long asymmetric strips as basic sensor elements. The strips usually extend the full length of the sensor, therefore giving only one-dimensional information on the location of the hit on the sensor. The biasing of the strips is usually fed from a shared connection line. The width is in the order of tens of microns, while the pitch between strips can be from 50 μm up to several hundreds of microns. The much lower density of sensor elements compared to pixel sensors, significantly reduces the number of readout channels. This makes strip sensors comparatively cheaper in terms

2. Silicon Strip Sensors

of costs per sensor area, therefore making it the optimal choice for the large areas of the outer tracking layers as in the CMS tracker (see section 3.1.2).

The following section will review the design basics of standard silicon strip sensors. A 3D-model of the full layout as used in the CMS experiment is shown in figure 2.3.

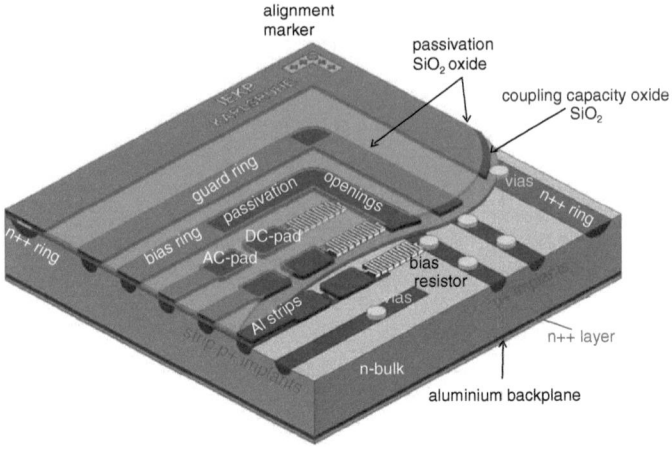

Figure 2.3.: 3D-model of a standard silicon strip sensor as used in the CMS experiment. It features AC-coupled readout strips with polysilicon biasing and no intermediate strips and the active area is protected by a single guard ring structure. Illustration kindly taken from [14].

2.2.1. Strip Geometry

For practical reasons, only the upper electrode of a strip sensor is segmented into strips, while the lower electrode called backplane covers the full backside of sensor. The readout electronics is connected to each strip and collects the charges generated by incident particles. The strip pitch depends on the requirements of the spatial resolution and the density of readout channels. The width of the strips, or - more importantly - the ratio between width and pitch (width-to-pitch ratio r_{wp}) is driven by the balance of low capacitance between the strips (low r_{wp}) and the need to reduce the field peak at the implant edges.

The usage of intermediate strips which are not read out, can increase the spatial resolution by capacitive charge sharing between adjacent strips, while keeping the density of readout channels constant. To take advantage of such a strip geometry, the readout has to measure the

2.2. Design Basics of a Silicon Strip Sensor

height of the signal which is generally referred to as *analogue readout*. In comparison, a *digital readout* only checks, if the signal exceeds a certain threshold while discarding the actual signal height. Therefore such a technology cannot take advantage of interpolation algorithms based on charge sharing.

2.2.2. DC to AC coupled Strips

As described in section 1.5.1 a p-n junction in reverse bias operation still shows a small reverse bias current. In an actual sensor, each strip exhibits such a small DC current in the order of nA. In heavily irradiated strip sensors, the reverse bias current can rise significantly by several orders of magnitude reaching a few μA per strip.

The amplifiers in the readout chips have to be protected from this current using current compensation circuits. These high pass filters remove the DC fraction of the current, allowing only the AC signals to pass to the amplifier. This functionality can also be incorporated into the sensor using large capacitors between the implant and the readout connection for the chips.

The readout capacitors are created by covering the strips with an aluminum electrode of about equal size above the strip implant, separated by a thin layer of oxide. The capacitance of this parallel-plate capacitor can be maximized by minimizing the thickness of the oxide according to $C = \varepsilon \cdot A/d$. It offers good filtering of DC currents, while having a strong coupling of the AC signals to the readout electrode which are connected to the amplifiers of the readout chip.

While DC coupled strips can be biased from the readout electronics, a separate biasing scheme is required for AC coupled sensors.

2.2.3. Biasing of the Strips

Each strip needs a reverse bias voltage to deplete the underlying sensor bulk from charge carriers. An aluminium metallisation covering the full backside of the sensor will serve as contact to the bulk. Each of the strip implants themselves needs to be connected to the other polarity of the reverse bias voltage. In DC coupled sensors this can be provided by the readout electronics which are directly connected to the implant.

In AC coupled devices[1] the biasing of the strips has to be done by additional structures on the sensor. For practical reasons, they are connected to a ring surrounding the sensor called

---

[1] Many DC-coupled device implement a form of external biasing from a common bias connection as well to allow the electric testing of the device. Otherwise each of the strips would need an individual connection to the bias voltage. Furthermore strips which are not properly connected to the readout electronics in the final detector cannot be properly depleted without a common bias scheme.

2. Silicon Strip Sensors

bias line, which provides the same potential to all of the strips from a single connection to the external voltage supply.

This way all of the strips would be shorted together. To electrically isolate the strips from each other, each strip is connected to the bias ring using a resistor. Additionally they also serve as current protection for each strip. These resistors can be implemented in various ways (all pictures from [14]):

Polysilicon resistor A resistor made of doped polysilicon is embedded in the sensor. It has a linear ohmic behaviour and is very radiation hard. The additional deposition and structuring of polysilicon requires additional process steps during manufacturing of the sensor, making the device more expensive.

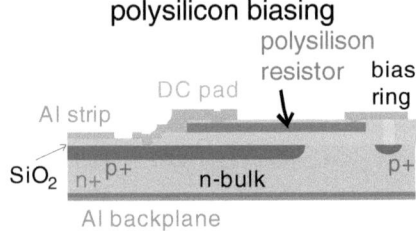

FOXFET A small gap between the strip implant and the bias ring is covered by a metal-oxide structure. Similar to a <u>M</u>etal-<u>O</u>xide <u>S</u>emiconductor <u>F</u>ield <u>E</u>ffect <u>T</u>ransistor (**MOSFET**) the effective resistance of the gap can be controlled using a small voltage applied to the metal gate. This structure is often referred to as FOXFET, as it utilizes the thick <u>F</u>ield <u>OX</u>ide as gate insulator. Cheaper but less radiation hard than polysilicon resistors.

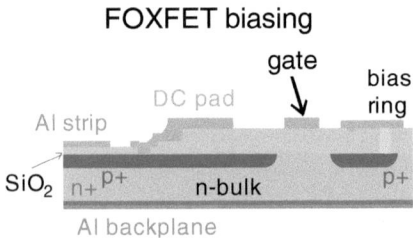

Punch through Very similar to a FOXFET but without a gate and a smaller gap. The size of the gap and the electric properties of the oxide define the resistance of the structure. Cheaper but less radiation hard than polysilicon resistors.

2.2. Design Basics of a Silicon Strip Sensor

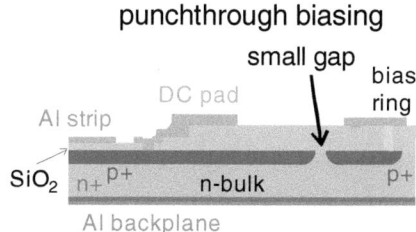

2.2.4. Breakdown Protection

Strip sensors are usually operated at full depletion. As discussed in section 1.6, the full depletion voltage V_{FD} changes with irradiation. Depending on the technology and material used for the sensor, it may even become inoperable if it is not fully depleted. For heavily irradiated sensors, V_{FD} can reach several hundreds of volts. Such high operation voltages may induce breakdowns, making the sensor inoperable or even destroying it.

Several methods exist to protect the sensor from breakdown. We present the most commonly used techniques used in the strip area and towards the edge of the sensor.

Guard Ring

The bias voltage supplied to the bias ring will drop to the backside potential in the small area between the edge of the ring implant and the sensors cutting edge. This will introduce high field strengths in an area afflicted by strong lattice deformations and contamination due to the cutting.

To lessen the effect of the abrupt change in potential, an additional ring of implant connected to the aluminium above it surrounds the bias ring. These so-called *guard rings* are floating and ensure a defined drop of the bias voltage along a larger distance. A *multi guard ring* structure implements several concentric rings, ensuring an even wider spread of the voltage drop in small steps. They are usually all on a floating potential were the potential drop from one ring to the next can be adjusted via a punch trough biasing.

Edge Ring

As mentioned before, cutting the sensor from the wafer introduces stress to the crystal lattice and increases the concentration of impurities. This produces a large dark current coming from the edge of which the sensor needs to be protected. An highly doped implant of opposite type than strips and the rings can protect the sensor from such edge effects as the space charge

2. Silicon Strip Sensors

region cannot reach the cutting edge. The high doping makes the material low resistive, thereby ensuring a more homogeneous field distribution.

Strip Area

The gap between the strip implants defines the field configuration in this area. In an under-depleted sensor with a narrow gap situation (see figure 2.4), the charge carriers located under the MOS structure in the interstrip region will get trapped when the depleted area separates them from the undepleted rest of the bulk. For a wide gap situation, these oxide surface charges are removed and the space charge region reaches the oxide as seen in figure 2.5.

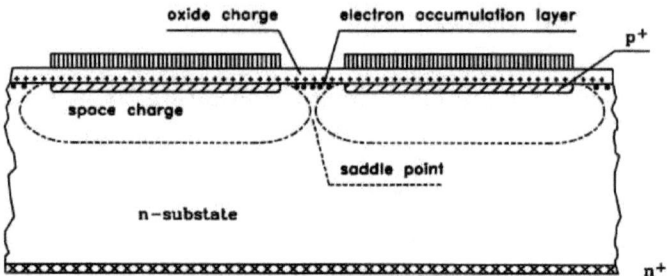

Figure 2.4.: Narrow-gap situation in an p-on-n sensor. The electron accumulation layer below the oxide will be trapped by the space charge regions joining from the two adjacent strips or p-n junctions respectively. The under-depleted region will influence the electric field distribution between the strips

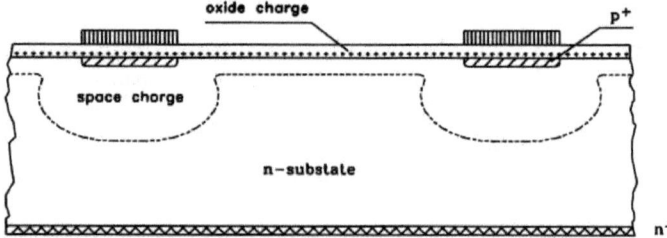

Figure 2.5.: Wide-gap situation in an p-on-n sensor. The electrons in the accumulation layer are depleted before the space charge regions created from adjacent strips or p-n junction respectively. As soon as the accumulation layer is gone, the space charge region will grow from the oxide towards the backside and the bulk is therefore depleted up to the oxide-silicon interface.

The electric field configurations are different for these two cases. While the maximum is at the edges of the implant for both situations, the narrow gap shows a significantly lower value as seen in the simulations shown in figure 2.6. To ensure a stable operation of the device at high bias voltages, the narrow gap solution is favored while it has to be balanced with the higher capacitance between the strips.

To further increase the high voltage stability of strip sensors and make them operable even after high irradiation, so-called *metal overhangs* where introduced. The metal covering the implants in strips and bias, guard and edge ring is larger by a few microns, than the implant itself. This extends the electric field lines into the thick oxide between strips and rings, reducing the field strength by a certain fraction. More important is the much higher breakdown voltage in oxide ($\approx 10^7 V/cm$ for SiO_2, see [2]) compared to bare silicon ($\approx 3 \times 10^5 V/cm$, see table 1.1).

An additional method to improve the high voltage stability of the sensors and indeed a very obvious one and easy to implement, is the avoidance of any sharp corners in the design. This reduces the risk of creating strong electric fields which encourage breakdowns.

2.2.5. Contact Pads

Contact pads are used to gain electrical access to the active structures. Such pads are used to supply the bias voltage, connect the readout strips to the readout chip through wire or bump bonding and to place probe needles for test purposes.

In the case of AC coupled sensors, two different kinds of contact pads are used:

DC pads are used to contact the implant of the strip and are usually used to perform electrical tests. They can only be placed at either end of the strip, as the contact pad and readout metallisation are placed in the same aluminium layer and have to be separated. The size of the pads is usually kept small, as the readout metallisation should extend over us much of the implant as possible. Strips with polysilicon resistor biasing usually require a DC pad to connect each strip to the bias ring due to the manufacturing process.

AC pads are used to contact the readout metallisation of the strip. Such pads are used to connect each strip to the input of the readout chip. These pads can be freely distributed along the strip which enables them to be larger. Additionally, most designs incorporate several AC pads per strip, to enable multiple contacts with bonds or needles.

2. Silicon Strip Sensors

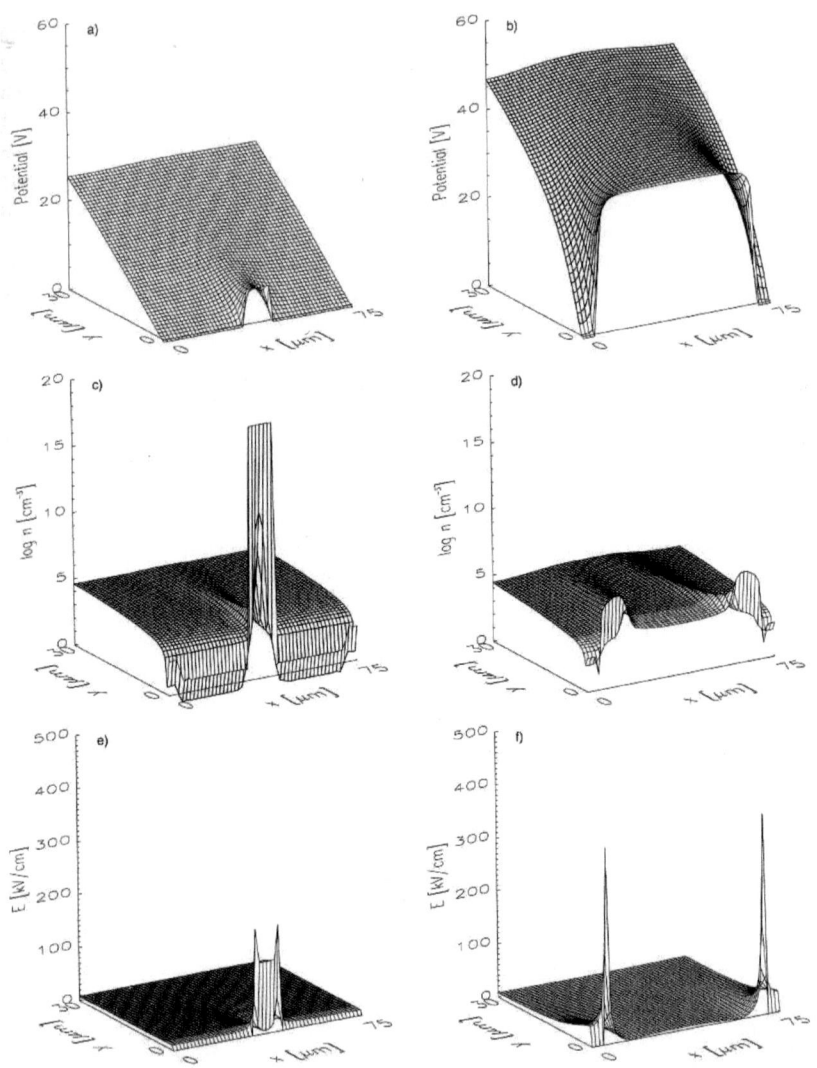

Figure 2.6.: Potential, electron density and electric field of a capacitively coupled strip detector in a narrow-gap (10 μm, left column) and wide-gap (65 μm, right column) situation. A strip pitch of 75 μm, a detector thickness of 300 μm, a bulk doping of $N_D = 2 \times 10^{12} cm^{-3}$, an oxide charge density of $N_{OX} = 3 \times 10^{12} cm^{-2}$, an oxide thickness of $d_{OX} = 220 nm$ and a reverse bias voltage of V = 130 V are assumed. The region from the center of one strip to the next is shown. Diagrams are taken from [15]

2.3. Manufacturing of Silicon Sensors

The manufacturing process for silicon sensors largely follows well established production technologies used in the semiconductor industry. Nevertheless, certain requirements necessary for good quality silicon sensors are not common and will be pointed out in the following sections.

2.3.1. Silicon for Silicon Sensors

Silicon is abundant in nature in the form of quartz sand (SiO_2). Several physical and chemical processes are necessary to purify it to Electronic Grade Silicon (**EGS**) which has purity of 99.999999999% or better than 1/100 ppb.

The EGS has then to be transformed into a single crystal. Several crystal growth techniques are available but two methods are mainly used to grow macroscopic single crystal silicon for semiconductor applications:

Czochralski (CZ) Silicon Silicon is heated to only a few degrees above its melting point (1414°C) in a quartz crucible. A small single crystal serving as seed for the crystallisation process is dipped into the melt and slowly retracted from it. By carefully controlling the temperature of the melt and the pull rate and rotation of the seed crystal, a perfect single crystal is grown.

Float Zone (FZ) Silicon A polycrystalline silicon rod is brought into contact with a small single crystal seed. The rod is then heated locally using RF heating. Impurities tend to stay inside the locally melted silicon as their solubility is larger in the liquid than in the crystalline form of silicon. As the RF coil is moved along the rod, the silicon solidifies as perfect single crystal while the gathered impurities are moved towards the end of the rod. Crystal growth is again controlled by RF power and temperature, movement of the coil along the silicon and rotation speed of the rod.

To adjust the resistivity and type of the final crystal, dopants are added, either to the cover gas for FZ or directly into the melt for CZ production. Due to the additional removal of unwanted impurities, FZ silicon is usually more suitable for high resistivity applications like strip sensors, where the resistivity of the substrate determines the operating voltage of the detectors. Recent studies show an increased radiation hardness of CZ compared to FZ material, which is due to its higher intrinsic concentration of oxygen. More information on radiation hard silicon can be found in chapter 6.

The final single crystal ingot or boule (see figure 2.8) has a diameter of 4 to 12 inch and is then cut into wafers (see figure 2.9) with a thickness of several hundred microns. Companies

2. Silicon Strip Sensors

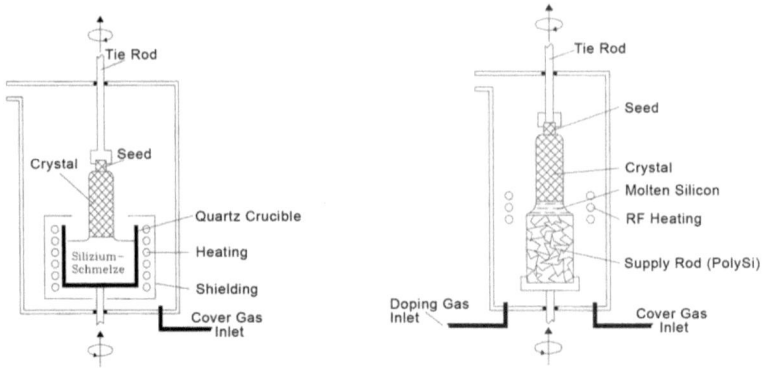

Figure 2.7.: Single crystal production using the Czochralski method (left) and the Float-zone method (right)[16].

and scientific institutes manufacturing silicon sensors are usually only capable of processing 4 or 6 inch wafers while in the traditional semiconductor industry 8 and even 12 inch wafers are standard.

Figure 2.8.: A single crystal silicon ingot or boule.

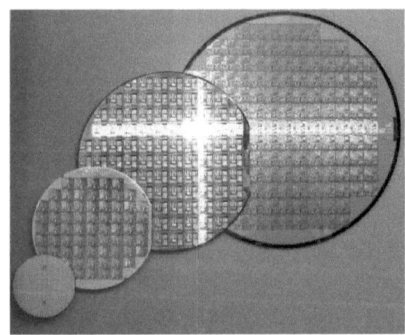

Figure 2.9.: Processed Wafers with 2 to 8 inch diameter.

Depending on the type and amount of dopands added to the crystal during production, the final wafers are either n- or p-type (e.g. arsenic or phosphorus for n-type and boron for p-type) with varying resistivity. Crystal orientation and thickness are further important parameters influencing the operating conditions of the final sensors (see chapter 1).

2.3.2. General Process Steps in Planar Technology

The manufacturing of silicon strip sensors uses the same basic process steps as in traditional semiconductor technology. The most important differences are stronger requirements in low contamination with impurities during production, especially in the oxides and uniformity over the wafer. Contamination of the sensor bulk will increase leakage current and change the operating voltage. An excess of charge carriers inside the oxides will deteriorating detector performance in radiation hard environments. The feature size on strip sensors is quite large, but the structures have to be free of defects over a very large area of a wafer compared to tiny structures on less then fingernail-sized areas in traditional semiconductors.

Nevertheless the basic production methods are very similar and will be described in the following subsections, concentrating on techniques which are commonly used for the production of silicon strip sensors. For a more complete description on silicon semiconductor technology see ref. [16].

Thermal Oxidation

One of the reasons for the success of silicon in the semiconductor industry, are the properties of its natural oxide. Silicon dioxide (SiO_2) or silica is mechanically hard and chemically inert, it has good electric properties like a dielectric constant of 3.9 and can be easily produced on silicon in very thin and uniform layers.

Thin layers of silica (a few nanometers to a micrometer) are formed by heating the silicon wafer in an oxygen enriched atmosphere. High Temperature Oxidation (**HTO**) at temperatures from 800°C to 1200°C can be performed using either molecular oxygen or water vapor and are consequently called dry or wet oxidation. The chemical reaction at the silicon to silicon oxide interface for the two oxidation methods is:

$$Si + O_2 \rightarrow SiO_2 \text{(dry oxidation)}$$
$$Si + 2H_2O \rightarrow SiO_2 + 2H2 \text{(wet oxidation)}$$

As the reaction consumes bare silicon it moves the interface deeper into the bulk of the wafer (see figure 2.10). For each unit thickness of silicon about 2.27 units of thickness of silicon dioxide are produced[16]. Therefore 56% of the oxide thickness will be above the original surface, while 44% will be below.

Oxygen has to diffuse through the already created silica layer which slows down the growth process with increasing thickness. Due to the higher solubility of water vapor (H_2O) than molecular oxygen (O_2) in silicon dioxide, wet oxidation is much faster. The growth rate is

2. Silicon Strip Sensors

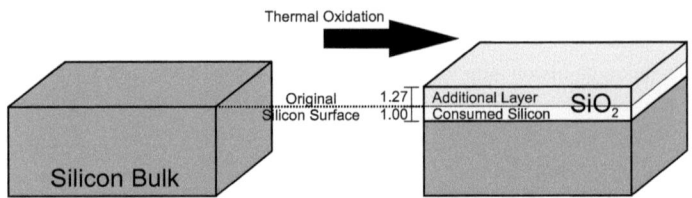

Figure 2.10.: Relocation of silicon surface due to oxide growth.

furthermore influenced by temperature, crystal orientation, pressure and additional additives to the process gas.

Unfortunately, the faster growth rate of wet oxidation reduces the quality of the oxide. Density and the dielectric strength are lower and the interface region to the silicon suffers from more dangling bonds.

Deposition

Various materials can be deposited on the silicon surface:

Metals usually aluminium alloys, are used to form electrodes for capacitive coupling, bonding pads for external connection, vias for layer interconnection and general routing of signals.

Polysilicon is deposited to create resistors and, which is rare for strip sensors, to serve as capacitive coupling electrodes or signal routing lines like metal despite its higher resistivity.

Silicon Oxide can be deposited at much lower temperatures (as low as 300 °C) than with HTO. Furthermore, deposition works on other substrate materials than silicon and is therefore necessary to create isolation layers on top of a metal layer. Nevertheless, the quality of the oxide is inferior to HTO. Silica is often doped with boron and/or phosphorus, e.g. BoroPhosphoSilicate Glass (**BPSG**), which makes them softer and they reflow to smooth out uneven surfaces.

Silicon Nitride (Si_3N_4) is used as a dielectric with a higher dielectric constant and electric strength than silica. It is very hard, dense and is a better diffusion barrier against water molecules than silica which explains it preferred use as a top layer passivation.

Several methods for depositing thin and uniform layers are in use:

Chemical Vapor Deposition (CVD) is primarily used to create layers of silicon nitride or oxide and polysilicon. The needed constituents which should be deposited, are added to a cover gas as chemical compounds. The compounds are decomposed by heating and deposited at the surface of the substrate. Several different implementations are used like Atmospheric Pressure CVD (**APCVD**), Low Pressure CVD (**LPCVD**) or Plasma Enhanced CVD (**PECVD**) which differ in the materials that can be deposited, the quality of the deposited film and the temperature needed to enable the reaction.

Epitaxy is the controlled growth of crystalline film on a substrate. It is often used to grow high quality layers of pure or doped silicon on a silicon substrate.

Sputtering is a physical deposition method. A target made of the material that is to be deposited on the substrate is bombarded with an ion beam. The high momentum ions knock atoms out of the target material and which are then deposited on the substrate. A gas which reacts chemically with the sputtered atoms of the target, can be added to the cover gas. The film deposited on the substrate consists of the product of the chemical reaction (e.g. Al_2O_3 from an Al Target). This process is called *reactive sputtering*.

Doping

To create the p-n junctions, the heart of any semiconductor device, the relationship of charge carriers in the top layers of the bulk silicon has to be manipulated by doping (see chapter 1). The dopands are brought into the silicon either by ion implantation or by diffusion from a gas or a layer of material containing the dopand like heavily doped polysilicon.

For implantation, the effect of channeling has to be taken into account: ions which enter the crystal perpendicular to the crystal face, along a major crystal axis, will encounter less scattering and travel much deeper into the crystal. To prevent channeling, the surface of the substrate is covered by a thin coating which deflects the ions off the perpendicular axis before entering the crystal. After implantation the dopands are not yet integrated into the crystal lattice making a high temperature treatment at around 900°C necessary to activate the dopands.

Doping by diffusion is usually done in two steps as well, a low temperature step (around 900°C) diffuse the dopands into a shallow surface layer of the silicon and a high temperature step (around 1100 - 1250°C) to drive the dopands into the crystal, shape the doping profile and activate them.

2. Silicon Strip Sensors

Photolithography

To transfer the functional structures that are designed using computer programs similar to CAD software, a process called photolithography is used. The designed structures are first transferred to a chrome film on a glass substrate called photomask. Figure 2.11 shows a screenshot of a photomask design.

Figure 2.11.: Screenshot of a digital design layout for photomasks. Each of the different colors represents a mask used in a separate step of the manufacturing process.

Figure 2.12.: The structures in the mask prevent exposure of the photoresist subsequently transferring the layout of the mask onto the photoresist.

The surface of the wafer is covered with a photoresist, a light sensitive material. To ensure a homogeneous and thin film over the whole surface of the wafer, the photoresist is applied using *spin coating*. The wafer is rotated around its center axis with 1000 to 5000 RPM while a small amount of photoresist is applied to the center of the wafer. The high rotation speed distributes the chemical liquid into a very homogeneous film of around 1 μm thickness in 30 to 60 seconds.

The photoresist is exposed to light while being shadowed by the photomask as sketched in figure 2.12. After developing the photoresist and removal of the non exposed parts, the functional structures are transferred to the pattern of the photoresist. This can be used as a mask against implantation or etching.

Etching

As the layers of materials can only be applied to the complete surface of the wafer, they have to be structured in a second process. This is usually done by etching away the unwanted material.

As the etchant will completely remove the full thickness of certain material layer, areas which should not be removed have to be protected by photoresist as described previously. The

2.3. Manufacturing of Silicon Sensors

pattern in the photoresist is then transferred into the underlying layer by etching the unprotected regions. By choosing chemicals which have a different etching rate for the topmost layer and the underlying material, the etching process slows down or stops after removing the top layer. The difference in etch rates for a given chemical is called *selectivity* (see figure 2.13).

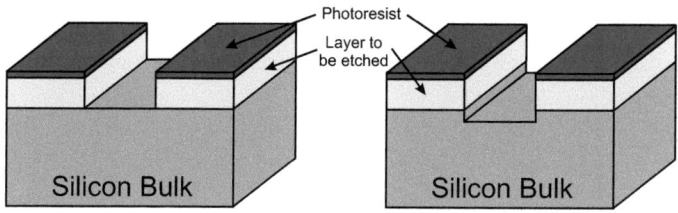

Figure 2.13.: Left: Good selectivity, only the layer to be etched is remove not the underlying material. Right: Poor selectivity, also the underlying layer is affected.

The etching rate can be spatially isotrop which leads to removal of material even below the areas protected by the photoresist. This so-called *underetching* has to be taken into account and corrected for in the design of the photomasks. Anisotropic etchants show different etching rates in certain directions relative to he substrates surface and can be used to avoid under-etching. This property of the etchant is called *isotropy* (see figure 2.14).

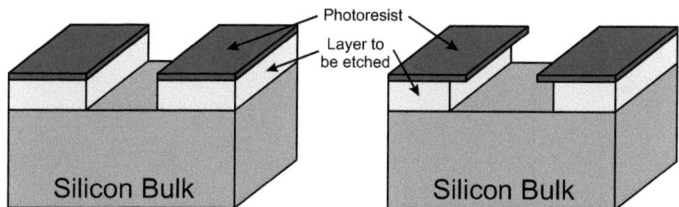

Figure 2.14.: Left: Highly anisotropic etching removes material only where it is unprotected by photoresist. Right: Isotropic etching removes material under the protective layer of photoresist as well.

The etching methods can generally be divided into two categories with several different varieties:

Wet Etching uses liquid etching chemicals to remove the solid material from the wafer and dissolve it in the alkaline or acidic solution. The wet etching solutions have a high selectivity but are in general isotropic. A few wet etchants show a high anisotropy for

65

2. Silicon Strip Sensors

the crystallographic axis of silicon. Due to the higher density in the (111) plane of the crystal, they have a significantly lower etching rate for these crystal planes.

Dry Etching methods allow good selectivity while enabling isotropic and anisotropic etching profiles. These methods rely on physical removal of material using an ion beam. Despite the high costs of the equipment, the versatility, high quality results and low consumption of chemicals has made dry etching the preferred method in the semiconductor industry. Nevertheless, the smaller companies and scientific institutes producing prototype or small scale productions of silicon strip sensors, still rely on wet etching. A variety of different methods exist like plasma etching, reactive ion etching or ion beam etching.

2.3.3. A Showcase Process Sequence

The processes described in the preceding sections have to be assembled into a process sequence to manufacture a working silicon strip sensors. The steps have to be arranged in such an order, that the sequence starts with the highest temperature process and ends with the lowest temperature ones. Otherwise a subsequent step at high temperature would influence or even damage the structures or materials formed in the preceding ones.

To better illustrate a possible sequence of process steps, a showcase implementation is presented here, using a not-to-scale but qualitatively correct pictures. The individual parameters for each process step are depending on the specific machines used by a manufacturer and are out of the scope of this work. It should be sufficient to understand the basic sequence of the necessary steps to correctly determine the final layout of a sensor from the design masks using the information represented in the following pictures. One has to keep in mind, that there are many additional steps necessary for the production which are not presented in detail here, like surface cleaning, fixing, developing and removing photoresists or additional high temperature treatments for dopand drive-in and diffusion.

2.3. Manufacturing of Silicon Sensors

Step 1: Plain silicon wafer

We start our showcase process implementation with a plain n-type silicon wafer. We cut the sensor along a strip from the edge of the sensor to slightly beyond the strips AC-pad. The situation is illustrated in the left picture.

Step 2: Oxidation and first photolithographic step

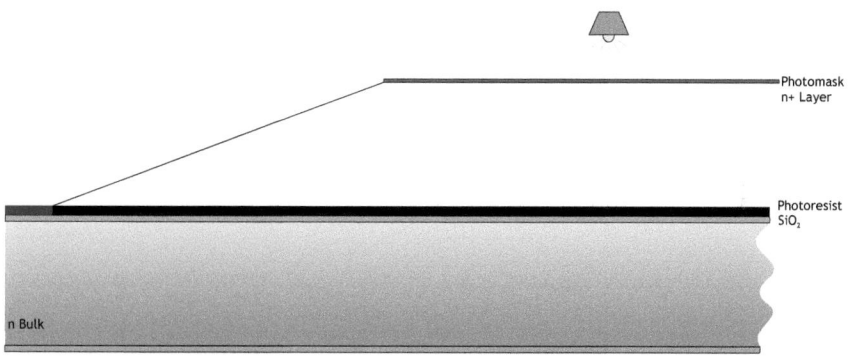

The first photolithographic step determines the layout of the n+ doping. A layer of (thermal) oxide is grown on top and bottom of the wafer and the photoresist is applied on top. The photolithographic step transfers the pattern of the n+ mask into the photoresist.

Step 3: Etching of oxide mask against n+ implantation

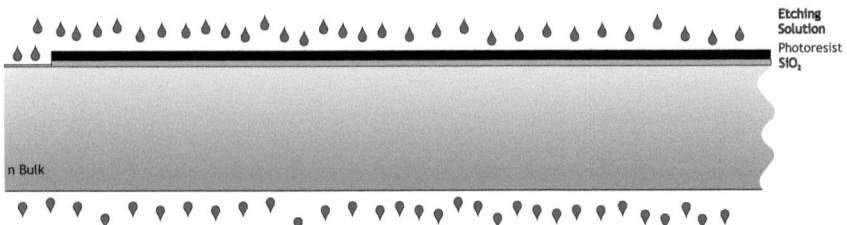

The pattern in the photoresist is now used to structure the oxide beneath by etching away the regions of the oxide uncovered by photoresist. The oxide grown on the backside of the wafer is removed in the same step as well.

Step 4: Phosphorus implantation/diffusion

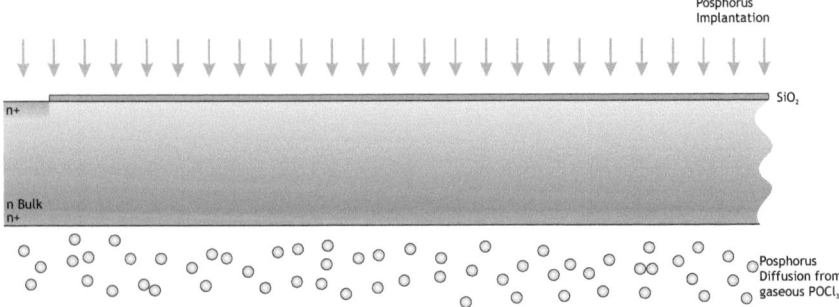

The unprotected regions of the bare silicon are now doped. Topside and backside can be treated with the same process step, or with separate methods by only exposing one side of the wafer. In the picture, the topside of the wafer is doped by implantation with phosphorus, while the backside is treated with diffusion from a phosphorous vapor to achieve a different characteristic. This step is followed by a thermal treatment to drive-in the dopands and activate them

2.3. Manufacturing of Silicon Sensors

Step 5: Oxidation and second photolithographic step

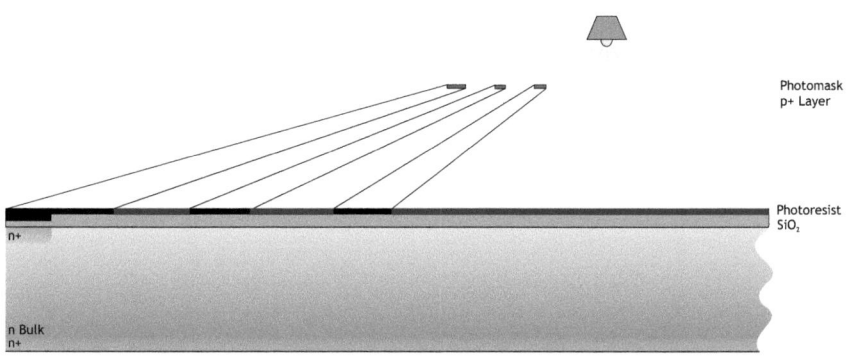

The next oxidation step and the application of photoresist on top is used for the photolithography of the p+ layer.

Step 6: Etching of oxide mask against p+ implantation

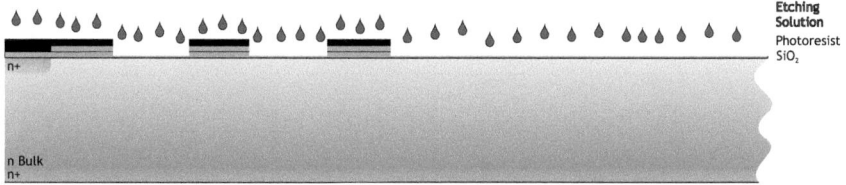

The photoresist is again used to transfer the pattern into the oxide beneath by etching. Areas which are not covered by photoresist are edged down to the bare silicon.

Step 7: Boron implantation

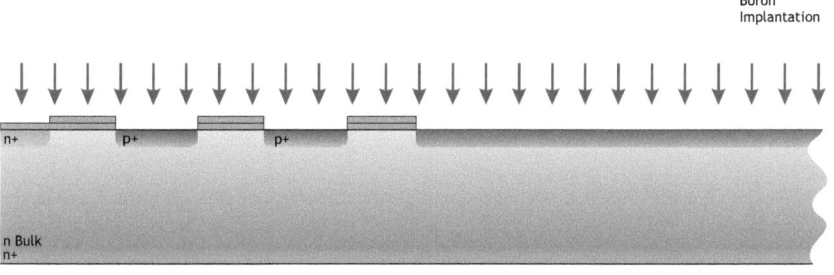

Boron implantation is used to form the p+ areas in the n-type silicon which defines the p-

2. Silicon Strip Sensors

n junctions. The actual strips and the implant of the bias and guard ring are formed in this process. The implantation is usually followed by a thermal treatment to drive-in and activate the dopands as in step 4. The diffusion of the dopands during this treatment slightly modifies the real location of the p-n junction with respect to the layout in the masks.

Step 8: Thin Readout Oxide (Gate Oxide)

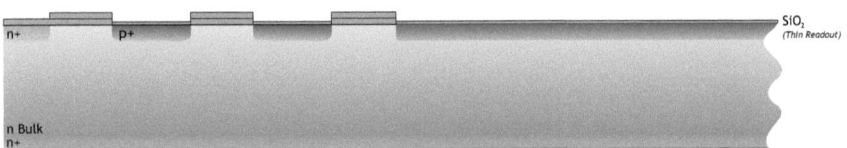

A thin layer of thermal oxide is grown on top of the p+ implanted areas. This oxide is called *readout oxide* as it forms the dielectric between the actual active detector strip and the readout metallisation which all together act like a capacitor. In the traditional semiconductor industry, this very thin layer of oxide is also referred to as *gate oxide*, as it similar to the oxide below the gate in a MOSFET transistor.

Step 9: Polysilicon deposition and doping

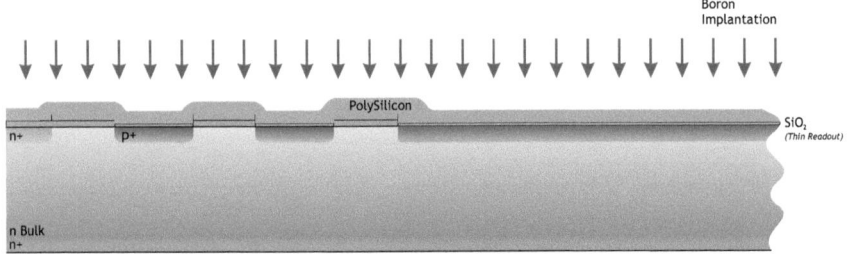

A layer of polysilicon is deposited on the topside of the wafer which will later form the bias resistors. The final resistivity of the polysilicon is adjusted by a subsequent implantation of boron.

2.3. Manufacturing of Silicon Sensors

Step 10: Third lithographic step

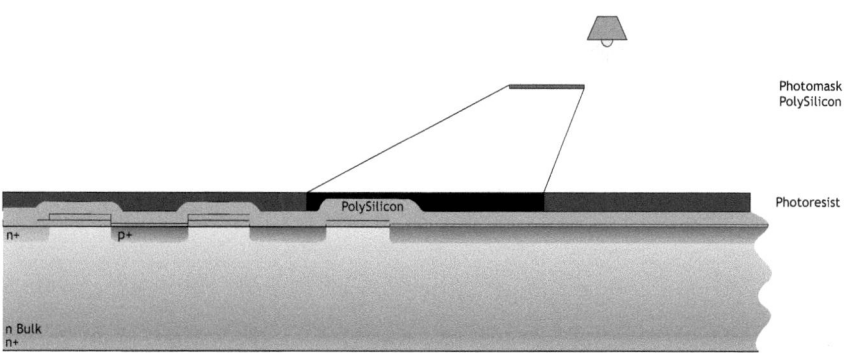

The layout of the polysilicon resistors is transferred into a photoresist applied on top of the polysilicon layer by photolithography.

Step 11: Etching of polysilicon

The formation of the bias resistors is done by etching off the polysilicon unprotected by the photoresist. The polysilicon is removed down to the underlying oxide layers.

Step 12: Fourth photolithographic step and implantation of the polysilicon heads

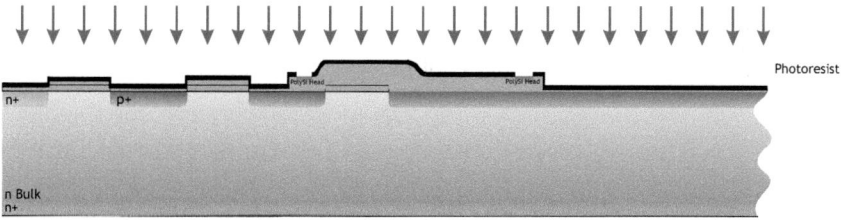

A layer of photoresist is again applied to the top of the full wafer and structured using a photolithographic step using a mask to form the polysilicon heads. The openings in the photore-

2. Silicon Strip Sensors

sists are sufficient to act as a mask against a boron implantation which forms small highly doped and low resistivity areas in the polysilicon. These areas improve the contact to the upcoming metallisation by preventing the formation of a Schottky Barrier (see section 1.5.3).

Step 13: Fifth photolithographic step

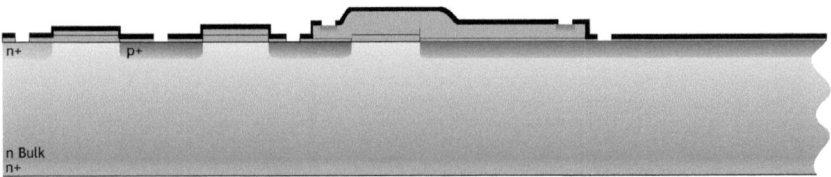

The next photolithographic step, again being a sequence of applying the photoresist, exposure to light through the photomask, removal of undeveloped photoresists and subsequent etching of the unprotected oxide, forms the vias. They provide a contact between the n+ and p+ doped silicon regions and the coming layer of metal.

Step 14: Deposition of metal and the sixth photolithographic step

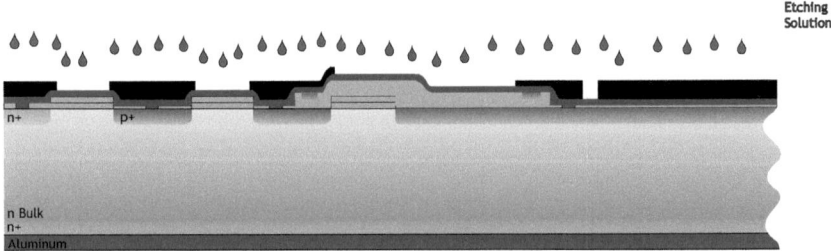

A layer of metal (usually pure aluminium or an aluminium alloy) is deposited covering the full surface of the wafer and structured with the same photolithographic steps as before. The patter in the mask used during this process provides the layout of readout strips, all the contact pads and the contacts for the rings surrounding the active area. The backside of the wafer is also fully covered by the aluminium layer, providing a good electrical contact to the n+ implanted backside and therefore the silicon bulk itself.

Step 15: Passivation and seventh photolithographic step

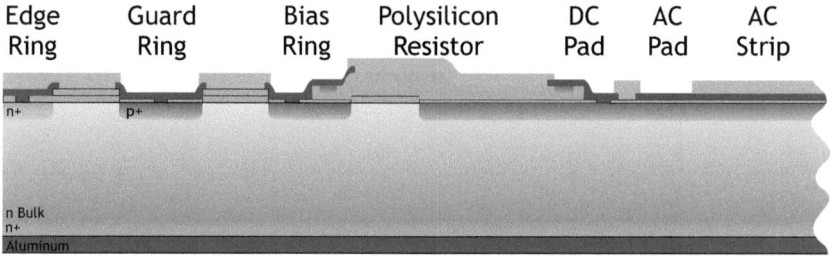

The final steps are used to apply a protection film against mechanical damage and humidity. This protective layer, also called passivation, is usually made of silicon nitride Si_3N_4 often interspersed with additional oxide layers. The layers are applied to the full frontside of the wafer and windows in the passivation are created with a photolithographic process. These windows are later used to access the metal contact pads to create the necessary electric contacts with probe needles or wire bonds. The design of the passivation mask is a little special, as the areas filled in the layout become the windows etched away from the oxide/nitride layer. To avoid confusion, this mask would better be called *passivation windows* instead.

The final structures are seen is this picture with the access pads contact the implants of the guard and the bias ring as well as the AC and DC pads to contact the readout strip and the strip implant respectively. The strip implant is connected to the bias ring *from the top* using the bias metallisation and the DC pad. This is necessary as it would be much more difficult to produce a good connection using a via between polysilicon and p+ implant.

The Compact Muon Solenoid (CMS) experiment is one of two large general-purpose particle physics detectors built on the proton-proton Large Hadron Collider (LHC) at CERN in Switzerland and France. Approximately 3,600 people from 183 scientific institutes, representing 38 countries form the CMS collaboration who built and now operate the detector. It is located in an underground cavern at Cessy in France, just across the border from Geneva.

Wikipedia on **Compact Muon Solenoid**

3

The CMS Experiment

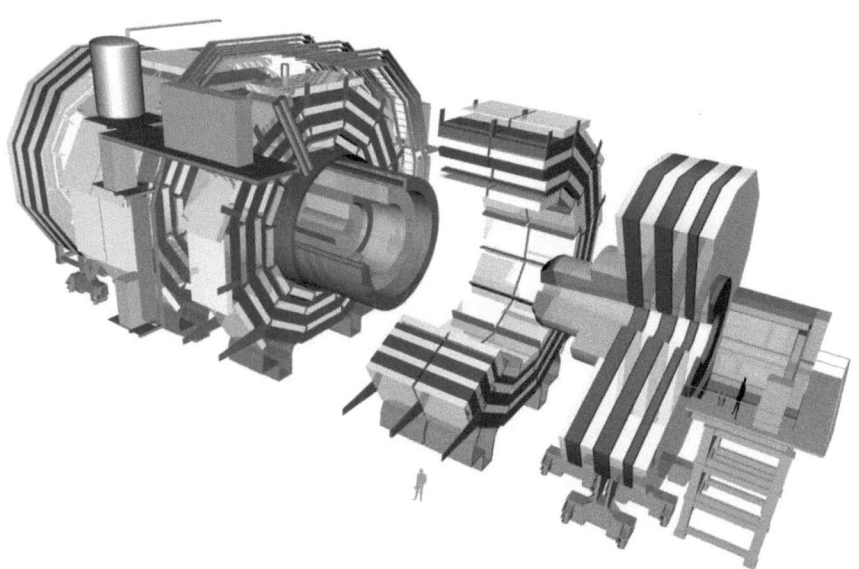

The Compact Muon Solenoid (**CMS**) is one of the four large experiments at the Large

3. The CMS Experiment

Hadron Collider (**LHC**). Similar to the ATLAS experiment[17], it was designed as a multi-purpose discovery detector assigned to tackle many of todays unanswered questions where the answers are accessible by the LHC for the first time. The main topics which drove the design choices of the CMS experiment are:

- Search for the Higgs Boson
- Search for supersymmetric particles
- Search for new massive vector bosons
- Search for extra dimensions
- Standard model measurements
- Heavy ion physics

To exhibit adequate performance on the mentioned activities, the requirements of the CMS detector can be summarised as follows:

- Good muon identification and momentum resolution over a wide range of momenta in the region $|\eta| < 2.5$, good dimuon mass resolution ($\approx 1\%$ at 100 GeV/c^2), and the ability to determine unambiguously the charge of muons with p < 1 TeV/c.

- Good charged particle momentum resolution and reconstruction efficiency in the inner tracker. Efficient triggering and offline tagging of τ's and b-jets, requiring pixel detectors close to the interaction region.

- Good electromagnetic energy resolution, good diphoton and dielectron mass resolution ($\approx 1\%$ at 100 GeV/c^2), wide geometric coverage ($|\eta| < 2.5$), measurement of the direction of photons and/or correct localization of the primary interaction vertex, π^0 rejection and efficient photon and lepton isolation at high luminosities.

- Good E_T^{miss} and dijet mass resolution, requiring hadron calorimeters with a large hermetic geometric coverage ($|\eta| < 5$) and with fine lateral segmentation ($\Delta\eta \times \Delta\phi < 0.1 \times 0.1$).

These requirements led to the some of the outstanding features of the CMS detector concept, like the high-field superconducting solenoid, the all silicon tracker and the fully active scintillating crystal electromagnetic calorimeter.

As most collider experiments, the general design of CMS resembles a barrel shaped onion: several cylindrical detector layers with discs covering both ends as seen in figure 3.1. The layered approach enables particles to be measured and identified using specialised detector systems at each of the layers. Although concentric spheres would be the optimal geometry,

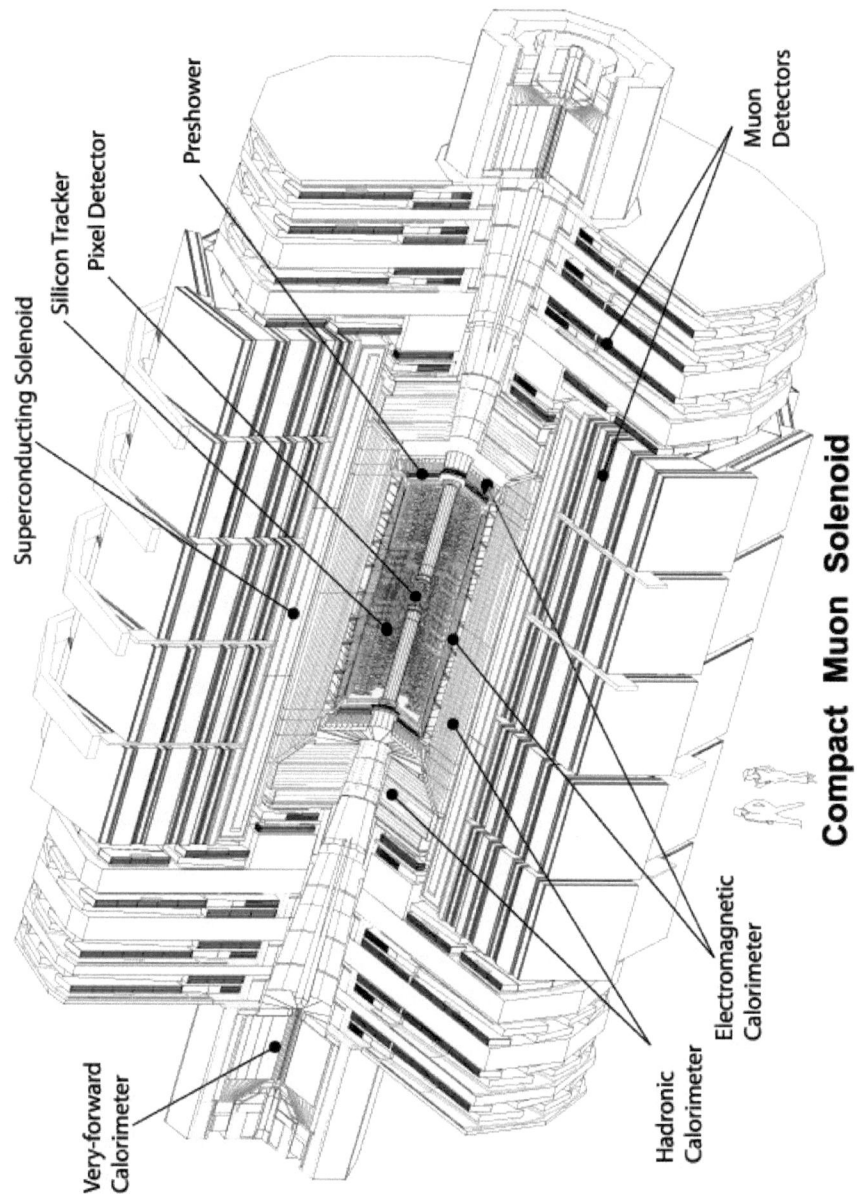

Figure 3.1.: 3D-view of the CMS experiment showing the location of the detector systems.

3. The CMS Experiment

giving the most homogeneous coverage, it is technically not feasible. Therefore the barrel-like shape proves to be the best compromise.

Figure 3.2 illustrates the path of the most important particles types when they traverse the detector. The task of the innermost detector system is to determine the particle's origin (vertex) and to measure the precise track of the particle, hence the name *tracker*. Thanks to the high magnetic field inside the tracker volume, the curvature of the particle tracks indicate the momentum and the sign of the electric charge of the incident particle. The calorimeters are measuring the kinetic energy of the particles and they are fully hermetic with only muons and neutrinos being able to pass through. The superconducting coil creates the magnetic field needed for the tracking system and the muon chambers for momentum measurement. The iron return yoke shapes the field in the outer part of the detector, where muon chambers determine the curved tracks of muons.

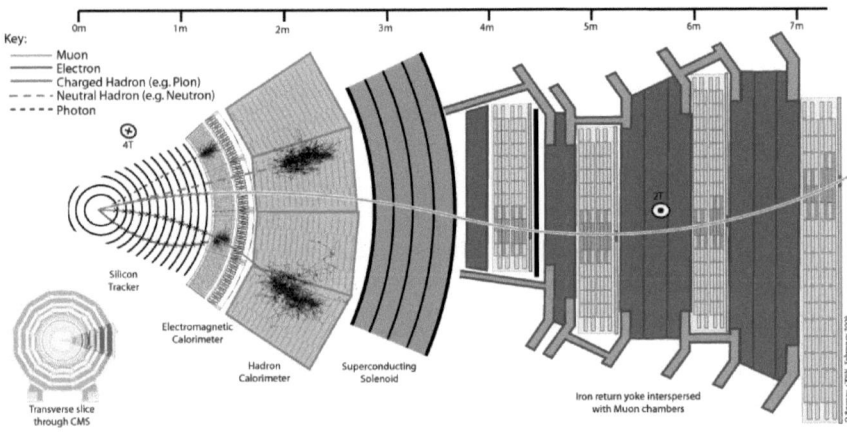

Figure 3.2.: Slice of the CMS detector showing the signals that different particles are inducing in the individual detector systems.

The CMS detector is 22 m long, has a diameter of 15m and a mass of 12,500 tons. Compared to the Eiffel Tower in Paris which has a total mass of 10,000 tons, the CMS experiment is a rather compact structure, hence the term compact in its name.

3.1. Tracking System

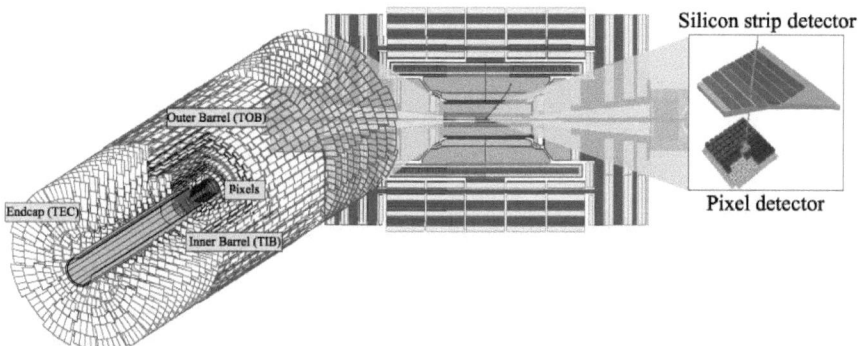

The first detector that particles are traversing is the full silicon tracking system. It consists of two different technologies, a pixel tracker and a strip tracker. While both give precise spatial information on particle tracks, they are optimised to satisfy slightly different demands[18].

3.1.1. Pixel Tracker

The pixel or vertex tracker is the innermost detector of CMS covering a mean radii of 4.4 cm to 10.2 cm from the beam pipe. Being so close to the interaction point exposes the detector and its accompanying electronics to high irradiation as seen in figure 3.3.

While having to cope with the high fluences, the pixel tracker should provide precise vertex position in the order of 10-20 µm. This is achieved by using silicon pixel sensors with 100 µm × 150 µm pixels and the ReadOut Chip (ROC) directly bump bonded onto it (see figure 3.4).

These detector modules are integrated into larger support structures. The full detector consists of three barrel layers with a total of 768 modules and four forward or endcap disks with a total of 672 modules (see figure 3.5). This results in over 65 million readout channels.

3.1.2. Strip Tracker

The Silicon Strip Tracker (SST) covers the mean radii from 20 cm to 110 cm. It comprises 15.148 detector modules with 24,244 silicon strip sensors each with 512 or 768 strips. This amounts to a total of 9,316,352 readout channels or a silicon area of 206 m^2 and makes it the largest silicon device today (2010).

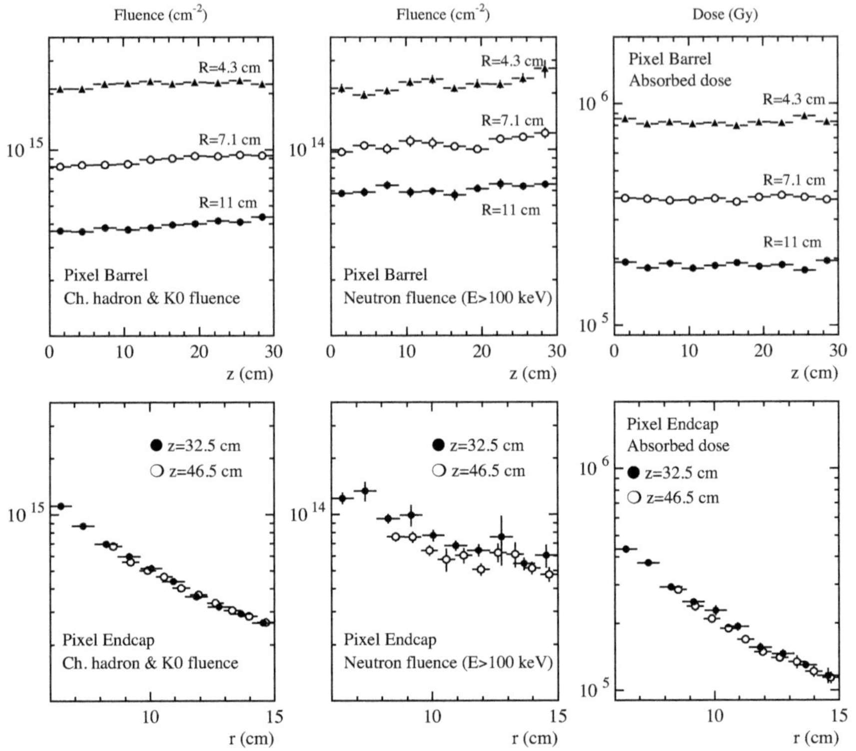

3. The CMS Experiment

Figure 3.3.: Simulation results of the energy-integrated charged hadron and neutron fluences and absorbed dose in the pixel detector. All values are for an integrated luminosity of 5×10^5 pb^{-1}[18].

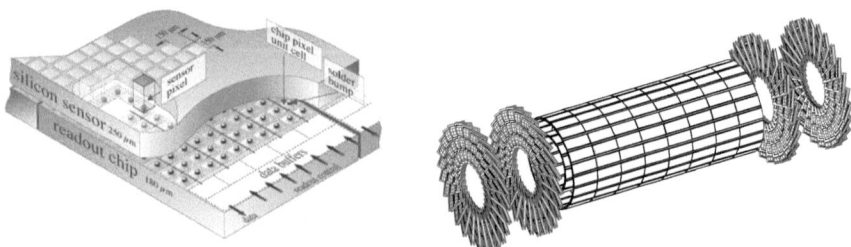

Figure 3.4.: Illustration of the CMS pixel sensor and the readout chip, which is directly bump bonded onto the sensor.

Figure 3.5.: The layout of the pixel detector with its barrel and endcap regions.

3.1. Tracking System

The SST is divided into four sub detectors, called the Tracker Inner Barrel (TIB), the Tracker Outer Barrel (TOB), the Tracker EndCaps (TEC) and the Tracker Inner Disks (TID).

The TIB covers the radii from 20 cm to 64 cm with four layers using sensors with a thickness of 320 μm and a strip pitch of 80 μm to 120 μm. The first two layers provide measurements in $r - \phi$ and $r - z$ simultaneously using two modules which are mounted back to back where one of the modules has a slightly rotated sensor (stereo module). The TOB consists of six layers and covers the radii up to 110 cm. Its sensors are 500 μm thick with a pitch of 120 to 180 μm. The thicker sensors should ensure a good signal-to-noise ratio as each module contains two daisy-chained sensors and therefore longer strips which exhibit higher noise. Again, the first two layers are equipped with stereo modules. The endcaps are divided into the smaller TID which covers the ends of the TIB and the TEC which extends to fully cover both ends of the tracker. The TID consists of 3 small disks with 3 rings of modules where the first two rings are again equipped with stereo modules. The two TECs are composed of 9 disks where each disk needs 12 different types of modules to fully cover the circular shaped area. The rings one, two and five are again equipped with stereo modules.

The full tracker provides coverage up to $|\eta| < 2.4$ while giving a minimum of 5 hits per track (see figure 3.6).

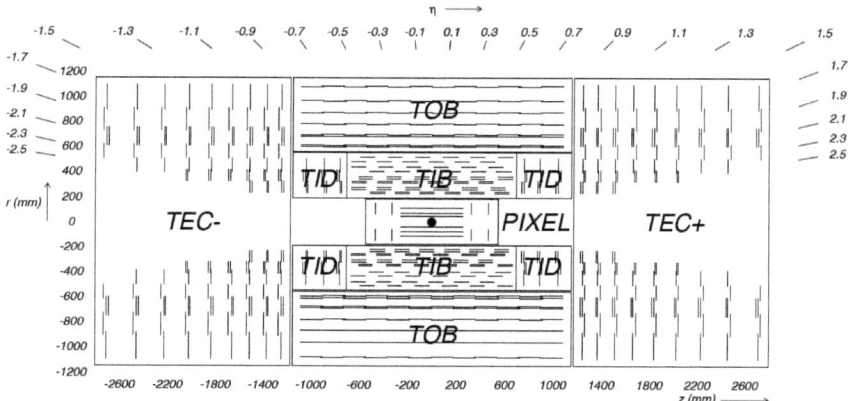

Figure 3.6.: Layout of the CMS tracker. This two-dimensional view is rotationally symmetric with respect to the beam running horizontally through the center. Each single line represents a sensor plane. Layers equipped with stereo modules and therefore providing two-dimensional hit coordinates, are easily spotted by two parallel lines close together.

3.2. Calorimeter System

The calorimeter systems measure the energy of the incident particles by fully stopping them in a dense material and measuring the deposited energy. The CMS detectors uses two different kinds of calorimeters, as it is common for most collider experiments - an electromagnetic calorimeter and a hadronic calorimeter. The former is placed immediately after the tracking system while the latter surrounds the electromagnetic calorimeter.

3.2.1. Electromagnetic Calorimeter

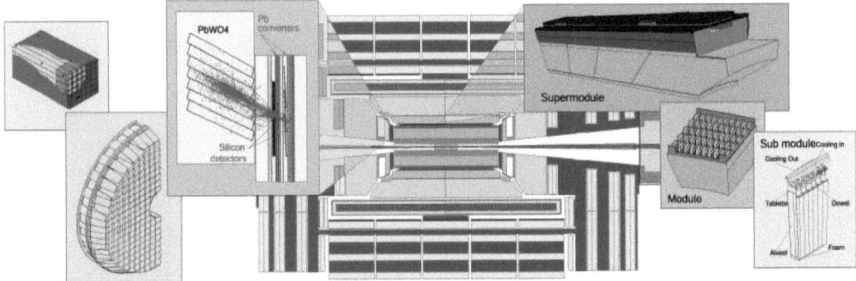

The Electromagnetic CALorimeter (**ECAL**) measures the energy of particles which are mainly interacting by the electromagnetic force like electrons, positrons and photons. It is a hermetic and homogeneous calorimeter made of more than 61,000 lead tungsten crystals (PbWO$_4$) in the barrel part and a further 14,648 crystals in the two endcaps[19].

Particles entering the crystals produce scintillating light by electromagnetic interaction. The lead tungsten crystals have excellent properties in terms of short radiation length (X_0 = 0.80 cm) and small Moliere radius (R_M = 2.2 cm), the light emittance is fast (80% within 25 ns) and the material is radiation hard. On the downside, the light yield is low (about 30 γ/MeV) which requires highly sensitive photodetectors.

To reliably detect the small amount of scintillating light produced inside the crystal, photodetectors with high intrinsic gain are used. The barrel section uses Avalanche Photo Diodes (**APD**) while in the forward parts Vacuum PhotoTriodes (**VPT**) provide the light measurement.

3.2.2. Hadronic Calorimeter

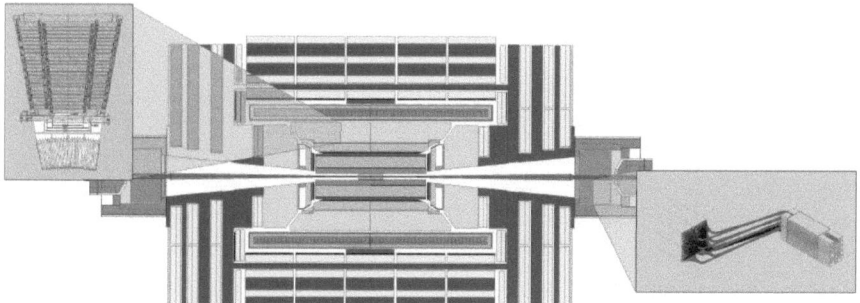

In contrast to the electronic calorimeter, the Hadronic CALorimeter (**HCAL**) measures the energy of the incident particle using strong interaction. CMS uses a sampling calorimeter, where plates of brass are used as absorber material interleaved with plastic scintillator as the active detector material[20].

A particle entering the hadronic calorimeter will interact with the nucleus of the absorber material creating hadronic showers. The particles in the shower will induce light in the plastic scintillators which is then transported by wavelength-shifting fibres to multi-channel Hybrid PhotoDiodes (**HPD**). The HPDs convert photons to electrons and direct them onto silicon sensors where these photoelectrons are detected.

Most of the HCAL material is inside the magnetic coil to minimize the non-Gaussian tails in the energy resolution and to provide a good containment and hermeticity for the measurement of the missing energy (E_T^{miss}). Only a small additional layer of scintillators, the Outer Hadron (HO) calorimeter, is placed beyond the magnet, lining the outside of the vacuum vessel to catch the tails still leaking through the calorimeters. This increases the effective thickness of the HCAL to over 10 interaction lengths.

Coverage for pseudorapidities of $3 < |\eta| < 5$ is maintained by the Hadron Forward (**HF**) calorimeter. Its main objective is to improve the measurement of the missing transverse energy and and to enable identification and reconstruction of very forward jets. As the available space in the forward region is much more cramped, the design had to be slightly different. To produce shorter and narrower showers the HF uses steel plates as absorbers and Cherenkov light emitted by quartz fibres embedded in the absorber plates as active detector element. The light is then transferred to photomultipliers for signal amplification and measurement.

3.3. Muon System

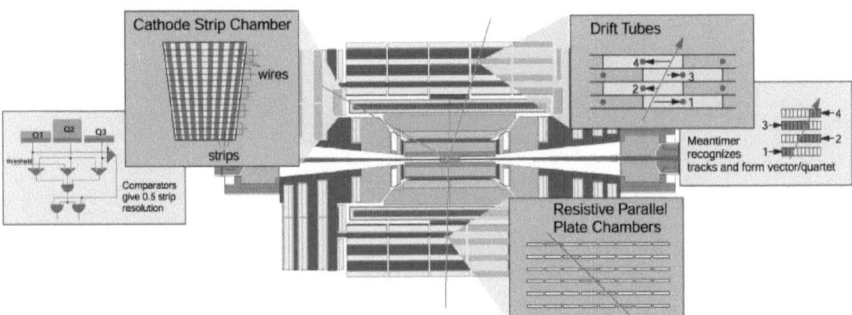

As the calorimeters are fully hermetic to hadrons and most leptons, the only particles that need to be measured past the HCAL are the muons. The muon systems should increase the momentum resolution for high momentum muons and provide fast information to the trigger. To achieve these tasks, three types of gaseous detectors are used for the measurement of muons[21].

Cathode Strip Chambers (**CSC**) are used in the endcaps, as the neutron background, the muon rate and the residual magnetic field is high in this region. For the barrel region, where the requirements are much lower, Drift Tube (**DT**) chambers are used. Both regions also employ Resistive Plate Chambers (**RPC**) to provide good timing resolution to identify the correct bunch crossing.

The CSCs consist of arrays of positively charged anode wires and and negatively charged cathode strips made from copper. The wires and strips are perpendicular and embedded in a gas volume. Muons, or any charged particle, ionize the gas and the electrons will move towards the anode wire while the ions will drift to the cathode strips. This gives spatial information in two coordinates for each passing muon and, because of the fast response of the closely space wires, the CSCs are also providing information to the trigger.

A 4 cm tube filled with a mixture of argon and CO_2 gas and a thin stretched wire placed along the center of the tube, are the basic components of the DTs. Electrons created by ionization from an incident particle drift towards the thin wire where they are registered. By calculating the distance of the ionization to the wire and the location of the induced signal along the wire, the DTs provide two coordinates for the position of the incident muon.

The RPCs can provide a high timing resolution in the order of nanoseconds. They consist of two parallel plates which are separated by a gas volume. These plates are made of a high resistive plastic material and are used as positively charged anode and negatively charged cathode. The electric field between the plates accelerates the electrons and ions created by ionizing particles towards the respective electrodes. The full charge created by the incident muon is collected by a metallisation at the backside of the plastic electrodes. As one of the metallisations is patterned into strips, the RPCs provides a spatial measurement which is used to estimate the muon momentum for the trigger.

3.4. Trigger System

The LHC provides collisions every 25 ns or at 40 MHz[1] where the full detector information of a single event amounts to approximately 1 MB. This would yield a sustained data rate of 40 TB/s which is not feasible to write to a storage system in terms of transfer speed and data capacity.

Most of the interactions do not involve any new physics processes, as the interaction cross-sections for interesting processes are rather small. Such events which show only well known physics can be discarded, leaving only those events for storage which might exhibit new physics. Nevertheless, this decision has to be made within a very small timeframe and at the high rate at which the LHC delivers the events.

The CMS experiment choose to implement a two level trigger system with a fully hardware implemented Level-1 trigger (**L1**) and an online filter system implemented in software running on a processor farm which is called High Level Trigger (**HLT**)[22].

The decision of the L1 trigger, whether to keep an event or discard it, is based on the presence of certain trigger primitives such as photons, electrons, muons and jets above certain E_T and p_T thresholds. This information is collected from the fast detectors in the calorimeters and the muon systems and only if the event is accepted by the L1, the complete detector information is read out and passed to the HLT. The L1 has a small time frame of 3.2 µs to make its decision which is based on the size of the buffers in the front-end electronics inside the detector . This timeframe is reduced by the transit time of the detector signals from the front-end electronics inside the detector to the service cavern housing the L1 trigger logic and the return back to initiate the readout of the full detector. The remaining time allocated for the L1 decision is less than 1 µs! To reach the required performance, while still retaining a high level of flexibility of the used algorithms, the L1 trigger logic is mainly implemented in Field Programmable Gate Arrays (**FPGA**). The achieved data reduction factor is in the order of $1:10^3$ resulting in a L1

---

[1]This is not entirely correct, as not all possible bunch locations inside he accelerator are filled. Certain locations are left vacant to allow the beam dump system enough time to ramp up the extraction magnets and safely extract the beam. Therefore the average collision rate is somewhat lower than 40 MHz.

3. The CMS Experiment

data rate of 100 kHz.

The HLT is implemented in software where an identical instance of the algorithms is run on several thousands of CPU cores of a computer farm. For each event accepted by the L1 trigger, the full detector information is passed to a processor of the HLT. A fast reconstruction of the event, similar to the full offline analysis but with less accuracy, is made to decide whether the event will be stored on disc or discarded.

A microstrip detector is a particle detector designed to consist of a large number of identical components laid out along one axis of a two-dimensional structure, generally by lithography. The idea is that several components will react to a single passing particle, allowing an accurate reconstruction of the particle's track.

Silicon microstrip detectors, in which the sensing mechanism is the production of electron-hole pairs in a 300-micrometre layer of silicon, with the electrons then being attracted by an electric field created by a pattern of interdigitated anodes and cathodes on the surface of the silicon separated by SiO2 insulator, are a common design.

Wikipedia on **Microstrip detector**

4

The CMS Silicon Strip Sensors

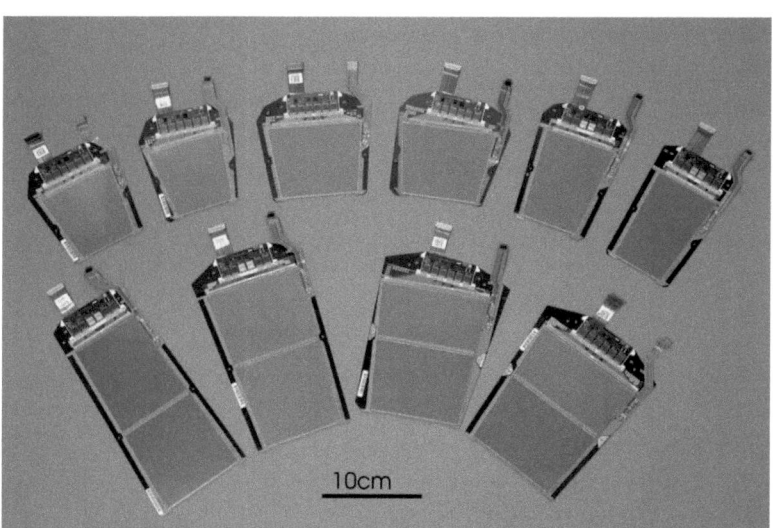

The following sections review the design choices that lead to the current sensor design of the CMS S̲ilicon S̲trip T̲racker (**SST**), the quality assurance scheme put in place to control the

4. The CMS Silicon Strip Sensors

production of more than 25,000 sensors and the design of the final detector modules. For a short review of the general layout of the SST and its four subdetector parts called TIB, TOB, TID and TEC, refer to section 3.1.2.

4.1. Sensor Design

The main sensor elements of the SST are roughly 10 cm × 10 cm single-sided, AC-coupled silicon strip sensors with polysilicon bias resistors and a single guard ring structure. It is manufactured in a standard p-on-n planar technology process on medium to high resistivity silicon. The details of the sensor material and design will be discussed in the following sections. For more information see the original investigation of design parameters [23], the invitation to tender[24] and the summary paper on the design and quality assurance[25].

4.1.1. Choice of Bulk Material

Following the results of the RD48 - ROSE Collaboration[26], standard float zone silicon was chosen as bulk material for the sensors. For the inner layers of the SST, which includes the TIB, the TID and the four inner rings of TECs (see section 3.1.2), thin sensors with a thickness of 320 μm provide a reasonable signal-to-noise ratio.

As the surface area A gets significantly larger with increasing radii r ($A = \pi r^2 z$) the number of readout channels would get unmanageable if the area covered by a single channel is the same as in the inner layers. To compensate that on the outer layers, spanning the TOB and the three outer rings of the TECs, two sensors had to be daisy-chained to create longer strips. This will increase the strip capacitance and therefore its noise. To compensate the increased noise, the thickness of the sensors in the outer layers was increased to 500 μm to again ensure a reasonable signal-to-noise ratio.

The thickness of the sensors has a direct influence on the full depletion voltage V_{FD} needed to remove all free charge carriers from the detector bulk, as described in section 1.5.1. Another important parameter which influences V_{FD} is the resistivity of the bulk material. This value is defined by the amount of dopands added during the manufacturing of the silicon crystal. The production of high resistivity materials gets increasingly difficult as the amount of unwanted impurities contaminating the silicon becomes the limiting factor.

Additionally, the resistivity changes with irradiation as described in section 1.6.2. For practical reasons, V_{FD} should be kept below a reasonable voltage which is determined by the capabilities of the high voltage power supply system. Using n-type silicon, V_{FD} will go down until type inversion after which it will rise as seen in figure 4.1. Having the V_{FD} at roughly the same

4.1. Sensor Design

voltage at the beginning and towards the end of operation maximises the available lifetime of the sensors. This led to the choice of 1.5 to 3 kΩcm for the thin sensors and 3.5 to 7.5 kΩcm for the thick sensors[24]. Due to difficulties in procuring materials of the above quoted values, the specifications for the thin sensors where later relaxed to 1.25 to 3.25 kΩcm. According to equation 1.65 the unirradiated depletion voltages are between 100 to 270 V.

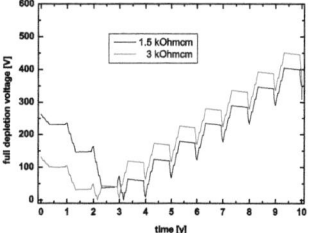 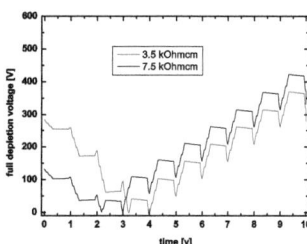

Figure 4.1.: Simulated 10 year LHC scenario calculating the evolution of the full depletion voltage of thin and thick sensors. V_{FD} is plotted versus the integrated irradiation dose in LHC running time. The model used to calculate the plot is based on the *Hamburg Model* using standard CMS annealing assumptions[27].

Furthermore, silicon with a lattice orientation of <100> was chosen instead of the more common <111> type. This reduces the number of dangling bonds at the silicon surface which in turn reduces the surface damage due to irradiation. It results in a slower increase in interstrip capacitance, which is the largest noise contribution in strip sensors operated at around -10°C.

4.1.2. Sensor Geometry

Sensors for the barrel regions are rectangular shaped while those for the endcaps have a more complicated wedge shaped geometry. One or two of these sensors are glued to a carbon frame together with pitch adapters, readout chips and some accompanying electronics.

Table 4.1 lists the geometries of the four sensor designs for the barrel region including the quantity required for SST. The rectangular sensors are of different size for the inner and outer regions, to better accommodate the requirements of their geometry, mainly the surface curvature due to the different radii. To balance the number of readout channels to the required two particle separation, each size comes with two different strip pitches. The pitch is tuned to match the modularity of the APV25 readout chips with 128 channels.

The design of the sensor geometries in the endcaps is driven by the effort to achieve a constant angular pitch with the strips pointing towards the interaction point. This results in

89

4. The CMS Silicon Strip Sensors

Type	Thickness (μm)	Length (mm)	Height (mm)	Pitch (μm)	Strips	Quantity
IB1	320	63.3	119.0	80	768	1536
IB2	320	63.3	119.0	120	512	1188
OB1	500	96.4	94.4	122	768	3360
OB2	500	96.4	94.4	122	512	7056

Table 4.1.: Sensor geometries for the barrel region of the SST[25]. IB1 will be mounted on 768 double-sided modules in the two inner layers of the TIB, IB2 in 1188 single modules in the two outer layers. OB1 will be mounted on 1680 single-sided modules in the layers 5 and 6 of the TOB, OB2 in the inner TOB layers (1-4) in single- and double-sided modules. All the TOB detectors are composed of two daisy-chained sensors.

wedge shaped sensors with different dimensions for each ring and a varying strip pitch in each sensor. The pitch is again tuned to the modularity of 128 readout channels. The eleven different geometries for the TEC and TID are listed in table 4.2.

Type	Thickness (μm)	Length 1 (mm)	Length 2 (mm)	Height (mm)	Pitch (μm)	Strips	Quantity
W1 TEC	320	64.6	87.9	87.2	81-112	768	288
W1 TID	320	63.6	93.8	112.9	80.5-119	768	288
W2	320	112.2	112.2	90.2	113-143	768	864
W3	320	64.9	83.0	112.7	123-158	512	880
W4	320	59.7	73.2	117.2	113-139	512	1008
W5a	500	98.9	112.3	84.0	126-142	768	1440
W5b	500	112.5	122.8	66.0	143-156	768	1440
W6a	500	86.1	97.4	99.0	163-185	512	1008
W6b	500	97.5	107.5	87.8	185-205	512	1008
W7a	500	74.0	82.9	109.8	140-156	512	1440
W7b	500	82.9	90.8	90.8	156-172	512	1440

Table 4.2.: Sensor geometries for the endcap region of the SST[25]. W1 has two different versions for TID and TEC, where as the TID shares identical W2 and W3 sensors with the TEC. The a and b type for W5 - W7 are used to daisy-chain two sensors and mount them on a single module.

All sensors feature the same basic layout of bias, guard and edge ring. A common bias ring surrounds the strip area to supply a ground connection to each strip. The ring has a metal overhang on the inside and the outside and small openings in the passivation at the corners to contact it. The bias ring is surrounded by a single guard ring with asymmetric metal overhangs. The edge is protected with a wide edge ring composed of n+ implant and a large metal overhang towards the wide gap to the guard ring.

The backside of all sensors is fully covered by aluminium, where a layer of uniformly n+

doped silicon is ensuring a good contact to the silicon bulk.

4.1.3. Strip Geometry

All strip sensors of the strip tracker use AC coupled strips with polysilicon resistor biasing. This enables the usage of the same readout chip throughout the SST. The strips have a constant width-to-pitch ratio of 0.25. This value is a compromise between a high ratio to reduces the field peak located at the p+ edge and a low value to reduces the total strip capacitance as explained in section 2.2.4.

The metal overhangs are in the order of 15% wider than the implant making them 4 to 8 μm depending on the sensor type. All structures have rounded corners to reduce the risk of high field concentration at sharp edges.

Polysilicon resistors where chosen for their high radiation hardness compared to other biasing techniques with a designated resistance of 1.5 (± 0.5) MΩ.

DC pads are placed at each end of the strip, where one of them is integrated in the connection of the resistor to the strip. These small pads are only used for the electrical test of the sensor. Two long AC pads are placed on each end of the strips to allow enough space for testing and the wire bonding of the readout chip in the final detector module.

4.2. Sensor Quality Assurance

Manufacturing a large quantity of silicon strip sensors, in the order of 25.000 pieces as for the CMS Tracker, has to involve an elaborate quality assurance scheme. During the production of the sensors, a certain fraction where electrically measured in dedicated probe stations. A full strip characterisation of a sensor, measuring several parameters on a per strip basis, took several hours and could therefore not be performed on all sensors.

A very useful tool to monitor the quality of the production with a reasonable effort, is the usage of test structures. Small areas on the wafer, which are not needed for the actual sensor, are used for dedicated structures measuring one or two important parameters. They can give access to values which cannot be extracted from a strip sensor, or involve a destructive measurement.

Assuming that production parameters are constant within one batch[1], it is sufficient to perform the measurements only on the test structures from one or two wafers among a single

---

[1] A batch or lot is a certain number of wafers in the order of 20-40 pieces, which run through the manufacturing process together. They are exposed to the individual processing steps together or consecutively with in short timeframe.

4. The CMS Silicon Strip Sensors

batch. If all parameters are within the specifications, the batch is assumed to be good. On the contrary, if any parameter is out of the specifications, a more thorough analysis is necessary determine the quality of the sensors.

In CMS, all wafers included the so-called *Halfmoon*, consisting of several test structures as seen in figure 4.2.

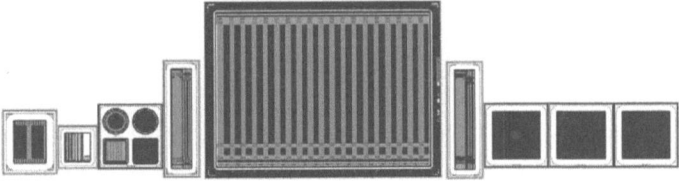

Figure 4.2.: The nine test structures of the CMS Halfmoon. The structures are called (from left to right): TS_Cap, sheet, GCD, Cap_TS_AC, Baby, Cap_TS_DC, diode, MOS1 and MOS2.

A closer description of each individual test structure and the measured parameters can be found in [28].

4.3. Detector Module Design

The bare silicon strip sensors are delicate devices which need a support structure for handling and mounting. Additionally, the small signals induced by incident particles have to be amplified as close as possible to the source to minimize signal loss and additional noise sources. This first stage of the electronic readout, together with the strip sensor and its support structure, is usually called a detector module.

CMS choose a carbon fibre frame as support structure for modules with a single sensor, while two sensor modules have an additional graphite crosspiece for mechanical enforcement as seen in the exploded view in figure 4.3 and in the picture in figure 4.5. These materials where chosen due to their high mechanical rigidity at low mass and a heat expansion coefficient similar to silicon.

A piece of Kapton foil electrically isolates the frame from the sensor and is also used to provide the high voltage to the backplane. A small board called *hybrid* is housing the necessary electric components and chips for the readout and monitoring of the module, mainly the APV25 readout chips (see [29] for a closer description of this important device). The inputs of the APV25 chips are routed to the bond pads of the sensor using a separate device called P̲itch A̲dapter (**PA**). It consists of a glass substrate on which aluminum lines are deposited. The pitch

4.3. Detector Module Design

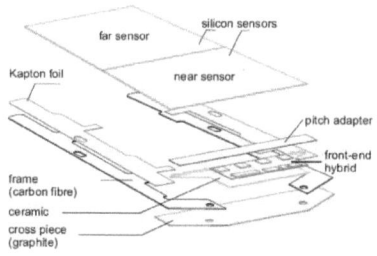 

Figure 4.3.: Exploded view of a two sensor detector module for the CMS tracker.

Figure 4.4.: Picture of a completed ring 2 module of the tracker endcaps.

of these lines fits the pattern of connection pads of the APV25 chip on one side and the sensors pads on the other side. The connection between PA to chip or sensor and the connection between the sensors on a two sensor module, are made using wire bonding. Figure 4.6 shows a picture of the hybrid with an already mounted PA.

 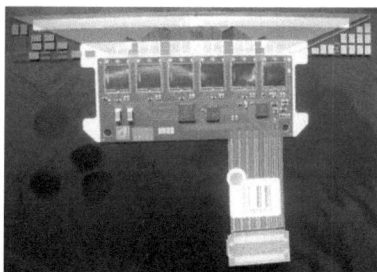

Figure 4.5.: Carbon fibre frame, kapton isolation and high voltage circuit for a ring 2 module on the assembly jig.

Figure 4.6.: The hybrid with its various readout components. The pitch adapter is already mounted.

93

Part II.

Sensor Design for the new CMS Tracker

Introduction

While the LHC experiments were only starting to collect their first data in 2010, the accelerator physicists and engineers were already planing for a possible upgrade to the LHC machine. An upgrade of the accelerator has great impact on the detectors as well and therefore the experiment collaborations were already starting their research and development on the implications and solutions to such an upgrade at the same time.

The first chapter of the second part discusses the possible upgrade paths of the accelerator and the impact on the experiments. It will show, that the CMS strip tracker will face a significantly increased radiation environment and it has to perform a new task in providing fast trigger information while a reduction in material budget is mandatory. The manufacturing of its components and the assembly of the complex detector should remain manageable by the participating institutes and fit within a tight financial budget and time schedule.

To overcome these challenges, new silicon strip sensors have to be developed using radiation hard materials and advanced layout techniques to produce the sensor and module design. The second chapter reviews the results from recent developments in radiation hard techniques made be the RD50 group at CERN. The third chapter describes the test structures for the evaluation of the production quality of such sensors, while the last chapter summarises the effort to design, produce and field test highly integrated sensors with added functionality.

The Super Large Hadron Collider (SLHC) is a proposed upgrade to the Large Hadron Collider to be made around 2012. The upgrade aims at increasing the luminosity of the machine by factor of 10 to 10^{35} $cm^{-2}s^{-1}$, providing a better chance to see rare processes and improving statistically marginal measurements.

Wikipedia on **Super Large Hadron Collider**

5
Super LHC and the CMS Upgrade

In 2010 the Large Hadron Collider (**LHC**) has successfully started with high energy collisions at 7 TeV center-of-mass energy. It will take a few more years until the accelerator will be

5. Super LHC and the CMS Upgrade

operated at the design energy and luminosity. Nevertheless the physicists and engineers of the accelerators and experiments started to think about a possible upgrade of the LHC machine and the impact on the experiments. This chapter summarises the motivations for the super-LHC (sLHC) and the possible upgrade scenarios. The consequences for the CMS Tracker which arise from the proposed scenario will be discussed as well.

5.1. Super LHC: a luminosity upgrade

The LHC accelerator was designed for a maximum luminosity of $10^{34} \text{cm}^{-2}\text{s}^{-1}$, which is expected to be reached gradually over the first years of operation. This would result in an integrated luminosity of about 100 $\text{fb}^{-1}/\text{year}$. The nominal operation time of the experiments is expected to be around 10 years.

After the initial ramp-up and a few years of running at a constant design luminosity the statistical errors will only improve very slowly with time, as evident from the simple estimations in figure 5.1. An increase of the luminosity of about one order of magnitude to $10^{35} \text{cm}^{-2}\text{s}^{-1}$ would be needed to significantly improve the measurements. The sLHC upgrade will be aimed at delivering this unprecedented luminosity at the same center-of-mass energy of 14 TeV. An energy upgrade is not feasible as it would require the exchange of all dipoles to achieve a higher magnetic field or use a tunnel with a larger circumference. Obviously, both options would result in a new accelerator than an upgrade of the already existing one and require enormous funds.

Assuming that the sLHC upgrade is technologically feasible for the accelerator and the experiments, the physics potential of such a machine could be divided into four main topics:

- Improvements of the accuracy in the determination of Standard Model parameters.
- Improvements of the accuracy in the determination of parameters of New Physics which is possibly discovered at the LHC.
- Extensions of the discovery reach in the high-mass region.
- Extension in the sensitivity to rare processes.

A more in-depth discussion on examples for the physics potential of the sLHC can be found in [30].

The first step in the increase of the luminosity has been planned since the design phase of the LHC and will not required any hardware changes to the machine. It will be accomplished by increasing the beam current at given emittance until the beam-beam limit reaches a value of 0.015 compared to the nominal value of 0.01. The corresponding luminosity was calculated as $2.3 \times 10^{34} \text{cm}^{-2}\text{s}^{-1}$ and is also called the *ultimate luminosity*.

5.1. Super LHC: a luminosity upgrade

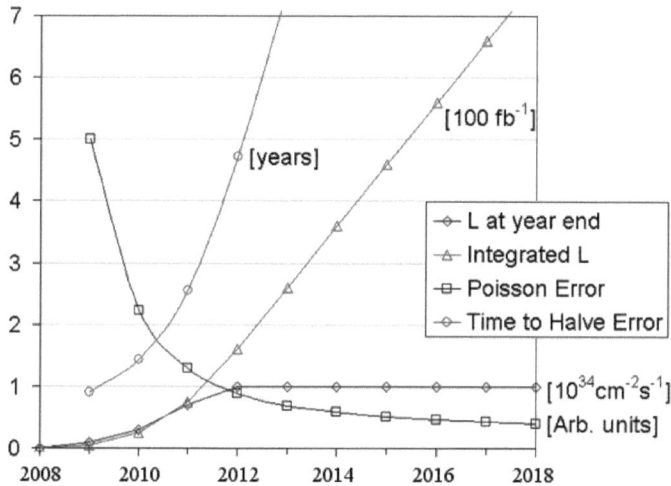

Figure 5.1.: Assuming the LHC will perform at design luminosity in 2012, the integrated luminosity will be given by the red line. The statistical error of any measurement will improve only very slowly after collecting the first few hundred fb^{-1} as shown by the black line or more drastically by the pink line, giving the time to half the error. The black line is given in arbitrary units and should just illustrate the evolvement of any statistical error.

To increase the luminosity further up to the sLHC value of 10^{35}cm^{-2}s^{-1} a major hardware upgrade of the machine is necessary. Several upgrade possibilities are discussed among the accelerator experts. The luminosity for a Gaussian beam is given by:

$$\mathcal{L} = \left(\frac{\beta_r \gamma_r f_{rev}}{4\pi}\right) \frac{k_b N_p}{\beta^*} \left[\left(\frac{N_p}{\varepsilon_N}\right) F(\Phi_p)\right], \qquad (5.1)$$

where β_r and γ_r are the relativistic factors, f_{rev} is the revolution frequency, k_b the number of bunches, N_p the number of protons per bunch, β^* the beta function at the interaction point ε_N the rms normalised transverse emittance and F the form factor. Furthermore, $\Phi_p = \theta_c \sigma_z/2\sigma^*$ denotes the *Piwinski angle*, $\sigma^* = \sqrt{\beta^* \varepsilon_N / \beta_r \gamma_r}$ the rms beam size at the interaction point, $d^* = \sqrt{\varepsilon_N / \beta_r \gamma_r \beta^*}$ the rms divergence at the interaction point and $\theta_c \approx ad^* \left(0.7 + 0.3b\sqrt{\tilde{k}_b \tilde{N}_p / \tilde{\varepsilon}_N}\right)$. In the last equation for θ_c the values for a and b are derived from simulations (see [31]) and the parameters with tilde are normalised to the the nominal values.

The proposed values for different upgrade paths are shown in table 5.2. At the time of writing, it has already been decided, that reducing the LHC bunch spacing to below 25 ns is not possible due to beam heating effects. A consequence of this is that the number of pp interactions in a bunch crossing becomes very high, 300 - 400 compared with $\approx$ 25 at the LHC. The 25 ns scenario results in greater peak luminosity, but because of its shorter lifetime, the average luminosity is similar to the 50 ns case. For physics measurements what is important is not the peak luminosity but the time integrated luminosity.

5.1. Super LHC: a luminosity upgrade

parameter	symbol	nominal	ultimate	more shorter bunches	bigger bunches	longer bunches	longer bunches II
emittance	ε [micron]	3.75	3.75	3.75	7.5	3.75	3.75
p+ per bunch	$N_b[10^{11}]$	1.15	1.7	1.7	3.4	6	4.9
bunch spacing	δt [ns]	25	25	12.5	25	75	50
number of bunches	n_b	2808	2808	5616	2808	936	1404
average current	I [A]	0.58	0.86	1.72	1.72	1	1.22
longit. profile		Gauss	Gauss	Gauss	Gauss	flat	flat
bunch length	σ_z [cm]	7.55	7.55	3.78	3.78	14.4	14.4 or 11.8
β^* at IP1&5	β^* [m]	0.55	0.5	0.25	0.25	0.25	0.25
crossing angle	θ_C [murad]	285	315	445	630	430	430
Piwinski parameter	θ_c	0.64	0.75	0.75	0.75	2.8	2.8
peak luminosity	$\mathcal{L}$ $[10^{34} cm^{-2} s^{-1}]$	1	2.3	9.2	9.2	8.9	8.9
events per crossing		19	44	88	176	510	340
IBS growth time	$\tau_{x,IBS}$ [h]	106	72	42	>100	44	54
nucl. scatt L lifetime	$\tau_N/1.54$ [h]	26.5	17	8.5	8.5	5.2	6.4
lumi lifetime (t_{gas}=85 h)	τ_L [h]	15.5	11.2	6.5	6.5	4.5	5.5
effective luminosity	$\mathcal{L}_{eff}$ $[10^{34} cm^{-2} s^{-1}]$	0.5	0.9	2.7	2.7	2.1	2.3
$T_{turnaround} = 10h$	$T_{run,opt}$ [h]	21.2	17	12	12	9.4	10.4
effective luminosity	$\mathcal{L}_{eff}$ $[10^{34} cm^{-2} s^{-1}]$	0.6	1.1	3.6	3.6	2.9	3.1
$T_{turnaround} = 5h$	$T_{run,opt}$ [h]	15	12	8.5	8.5	6.6	7.3
av. e-c heat load SEY=1.4 (1.3)	P [W/m]	1.07 (0.44)	1.04 (0.59)	13.34 (7.85)	2.56 (2.05)	0.26 (0.26)	0.36 (0.08)
SR heat load at 4.6-20 K	P^{SR} [W/m]	0.17	0.25	0.5	0.5	0.29	0.36
image currents power at 4.6-20 K	P^{IC} [W/m]	0.15	0.33	1.87	3.74	0.96	0.96
BG heat load for 100 h (10 h) LT	P^{gas} [W/m]	0.038 (0.38)	0.056 (0.56)	0.113 (1.13)	0.113 (1.13)	0.066 (0.66)	0.081 (0.81)

Table 5.2.: Suggested parameters for the LHC upgrades [32]. The ultimate values will be reachable without any hardware modifications to the machine, while the parameters in the rows to the right show possible scenarios for the sLHC upgrade.

5.2. Challenges for the Strip Tracker

At the time of writing, there are no physics justification for an increase in spatial or momentum precision in the tracker. The tracking and vertexing performance of the current tracker would be sufficient and have to be maintained at $10^{35}\text{cm}^{-2}\text{s}^{-1}$. The main challenge for the tracking system comes from the significant increase of interactions in each collision, producing many more particles within the same volume. The particles have to be detected and their tracks reconstructed while they are also irradiating and damaging the sensors and the electronics. A number of challenges are caused by this tenfold increase in luminosity which I will describe in this chapter, while the subsequent chapters will offer methods to tackle some of the presented challenges.

Furthermore, the material budget of a tracking detector has to be kept small, to minimise the influence on the particle tracks. As the current CMS Tracker already suffers from an excess of mass within the detector, the minimisation of the used material is imperative at every possible element. The solutions presented by me in this work are all geared towards a reduction of material compared to the concepts used in the current tracker.

5.2.1. Increase in granularity

As evident from figure 5.2, the track density within the tracker will increase significantly. The hits on each detector layer have to be associated to the track of the original particle which is performed by so-called pattern recognition algorithms during the off-line data reconstruction. Due to the higher density of tracks, the number of hits which are in close proximity will increase. This might ultimately lead to having two or more hits on a single strip. These individual hits of different particles cannot be separated and are only detected as one single hit. As a consequence, the performance of the pattern recognition will suffer as ambiguities in the combination of hits to tracks cannot be sufficiently resolved.

A sufficiently robust pattern recognition at sLHC luminosities can only be achieved by an increase in granularity of the strip tracker. Using sensors with shorter strips compared to the currently used designs, two tracks which would produce only a single hit on one long strip could be separated again. One has to keep in mind, that this does not effect the spatial resolution perpendicular to the strips but only along the strip.

Implementing strip sensors with such shorter, almost pixel like *strixels* of several millimetres to a few centimetres, produces additional challenges for the module design. Present connection methods, as used for the current tracker and described in section 4.3, will not be able to integrate sensors with several thousands of readout channels. The traditional interface between

5.2. Challenges for the Strip Tracker

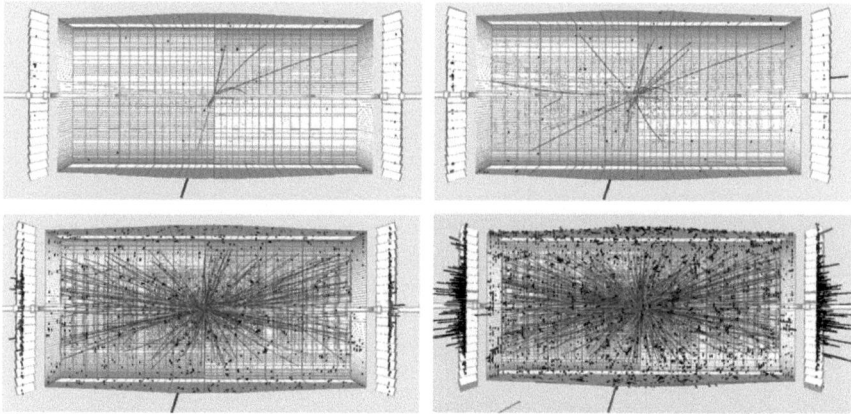

Figure 5.2.: Simulated event at various luminosities (clockwise form upper left: 10^{32}, 10^{33}, 10^{35} and $10^{34} cm^{-2} s^{-1}$).

readout chips and sensor using a separate pitch adapter and connecting the chips from the edge of the sensors reaches its limits in terms of readout channel density. One could utilise the remaining three edges of the sensor and connect additional pitch adapters and chips, but this would only shift the difficulties to the integration of modules with readout interfaces on all four sides into the trackers mechanical support. Furthermore, the maximum number of readout channels could only be increased by a factor of four.

A much more elegant approach would be the integration of the pitch adapters functionality on the sensor itself. An additional metal layer on top of the sensor could provide the routing from each strip to a connection pad, which can be directly bonded to the readout chip. This would not only make the material of the pitch adapters obsolete and reduce the number of bonds needed to connect each strip, but it would also offer a new freedom in the arrangement of readout chips. An advanced concept of such a module would have its highly integrated readout chips with several hundred to a thousand channels bump bonded directly on top of the sensors without the need of an additional circuit board or hybrid. Several chips could be used to read out the full sensor and only a small piece of Kapton would be needed to feed the power supply lines to the chip and the data from it. A sketch of the concept can be seen in figure 5.3. Apart from the proposed layout of the module, the integrated on sensor routing would offer the possibility for almost any arrangement of readout chips, enabling new possibilities for advanced module support or cooling concepts.

105

5. Super LHC and the CMS Upgrade

Figure 5.3.: Concept of a highly integrated module. The high density readout chips (gray) are directly bump bonded on top of the sensor where an additional metal layer integrated into the sensor connects the inputs of the chip to the sensor strips (yellow). The power lines and readout of the chip is supplied by a thin Kapton glued on top of the sensor (brown). The granularity of the sensor can be increased by integrating additional chips which are daisy chained to the same Kapton supply cable.

A very important step in the development of such highly integrated modules, is the implementation of on-sensor routing schemes and the successful integration into a strip sensor. The manufacturing process would have to be adapted to add the additional metal layer on top while careful attention has to be paid to any signal loss or cross-talk introduced by the routing.

An additional benefit of making the pitch adapters and hybrids obsolete and offering much more flexibility for the design of a sensor support, is the reduced complexity in the manufacturing of the module. Furthermore, fewer components means less material per module. This reduction in material within the tracker volume is very important, as one of the main criticisms of the current CMS tracking system is the material budget which is larger than it was anticipated in the first designs. This is shown in figure 5.4, where the left plot clearly shows, that secondary elements of the detector like cooling, cables and the support structure make a large fraction of the material budget.

5.2. Challenges for the Strip Tracker

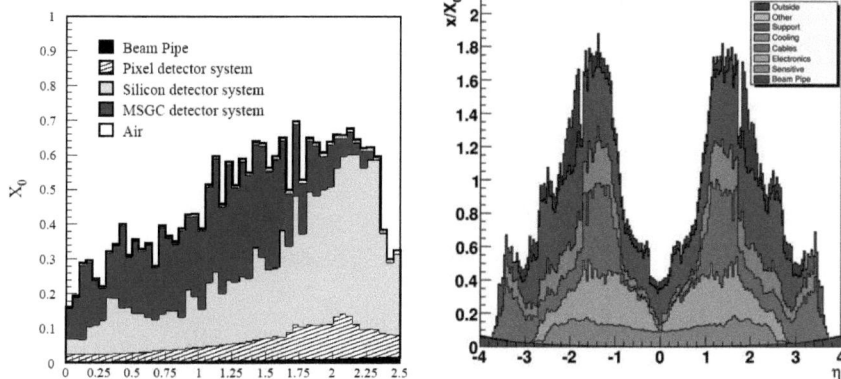

Figure 5.4.: Material budget as anticipated in early design reviews (with MSGC tracking) and in the actual all silicon tracker.

5.2.2. Increase in Radiation

The CMS Strip Tracker was designed to operate up to a integrated luminosity of approximately 500 pb^{-1} [18]. By the time the sLHC upgrade will become operational, the inner layers of the strip tracker will probably have reached the end of their lifetime. The full depletion voltage of the sensors in these layers will by higher than the maximum that the powers supplies can provide and because the sensors are already type inverted, it will be impossible to extract signals from the sensors.

The higher luminosity of the sLHC will increase the charged hadron and neutron flux inside the tracker significantly. It is assumed, that the integrated luminosity will be increased by a factor of 5 to 2500 pb^{-1}. The integrated fluence and dose at different radii from the interaction point are shown in table 5.4. Simulation results of the neutron flux and the absorbed dose are shown in figure 5.5

Comparing the CMS Tracker radiation environment for the LHC in table 5.3 and the sLHC in table 5.4, there is approximately a factor of five difference. It may be possible to use the same technology now used in the inner layers of the strip tracker for the outer layers in the upgrade, as they receive the same amount of irradiation. For the inner strip layers of the upgrade, current pixel technologies might be an option but they will probably be too expensive for larger areas, even in the coming years. Additionally, the granularity of a pixel detector is too fine resulting in too many readout channels which cannot be handled appropriately. To compensate for that, strip sensors using more radiation hard silicon bulk materials and process technologies have to

5. Super LHC and the CMS Upgrade

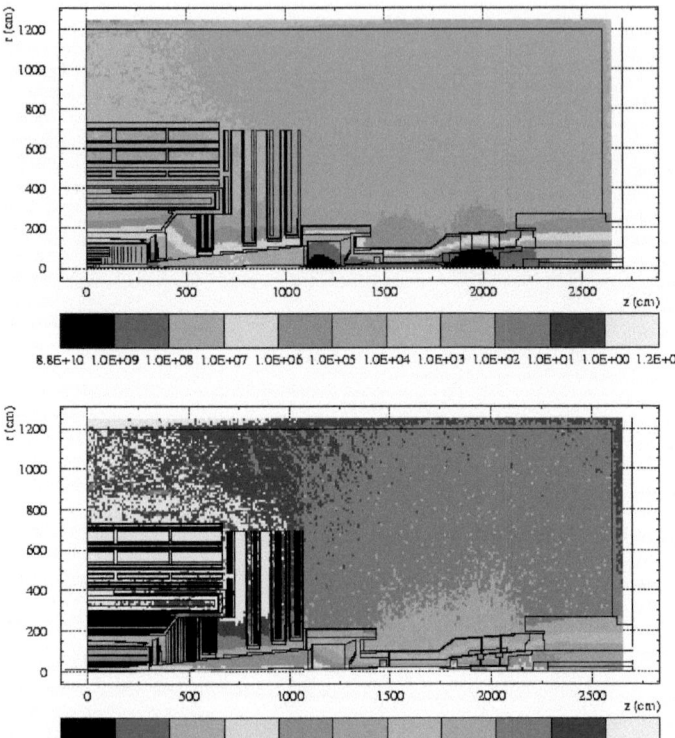

Figure 5.5.: Neutron flux ($cm^{-2}s^{-1}$) at an instantaneous luminosity of $10^{35} cm^{-2}s^{-1}$ (upper plot). Dose (Gy) for an integrated luminosity of 2500 fb^{-1} (lower plot). Plots taken from [30].

radius (cm)	fluence of fast hadrons ($10^{13} cm^{-2}$)	dose (kGy)
4.3	246	828
11	45.2	187
22	15.7	66.6
75	2.9	6.8
115	2.8	1.85

Table 5.3.: Hadron fluence and radiation dose in different layers of the barrel part of the CMS Tracker. Integrated luminosity is assumed to be 500 fb^{-1}. Data compiled from [18] and [33].

5.2. Challenges for the Strip Tracker

radius (cm)	fluence of fast hadrons ($10^{14} cm^{-2}$)	dose (kGy)	charged particle flux ($cm^{-2}s^{-1}$)
4	160	4200	5×10^8
11	23	940	10^8
22	8	350	3×10^7
75	1.5	35	3.5×10^6
115	1	9.3	1.5×10^6

Table 5.4.: Hadron fluence and radiation dose in different layers of the barrel part of the CMS Tracker. Integrated luminosity is assumed to be 2500 fb^{-1}. Data taken from [30].

be developed for the inner strip layers. The innermost part of the tracking system, currently equipped with pixel detector, will receive unprecedented irradiation in the sLHC environment and will demand very advanced or entirely new sensor concepts.

The fluence at the outer silicon tracker (r = 100 cm) will be about $10^{14} n_{eq}$ cm^{-2} increasing to $10^{16} n_{eq}$ cm^{-2} at the inner pixel (r = 4 cm). In the inner region (r < 20 cm) radiation is dominated by pions, while the outer region is dominated by neutrons with a considerable fraction of pions (> 25 %).

The main concerns for highly irradiated sensors is the leakage current and the signal-to-noise ratio. As explained in section 1.6, irradiation increases the leakage current of a sensor. It can increase by several orders of magnitude and can lead to the so-called *thermal runaway* where the high current increases the sensor temperature which in turn raises the leakage current. The most important feature of a sensor is its signal-to noise ratio where a minimum of 10 is favorable to have a reasonable efficiency in detecting hits and determining their position. Irradiated detectors will loose much of their signal, predominantly due to trapping in the detector bulk. Furthermore, traditional p-on-n sensors will only operate if they are fully depleted after type inversion, as the depletion layer is growing from the backside and the holes cannot reach the strip electrode if under-depleted. Therefore the full depletion voltage becomes an issue, where reverse bias voltages beyond full depletion are usually beneficial to the extracted signal as well. On the contrary, the current drawn by the module at higher reverse bias voltages will increase, again adding to the vicious circle of thermal runaway.

5.2.3. Tracking Trigger

The single muon, electron and jet trigger rates at the Level 1 Trigger (**L1**) will greatly exceed 100 kHz at sLHC conditions. It cannot be reduced sufficiently using higher p_T thresholds which becomes clear when extrapolating the L1 rates from figures 5.6 and 5.7.

5. Super LHC and the CMS Upgrade

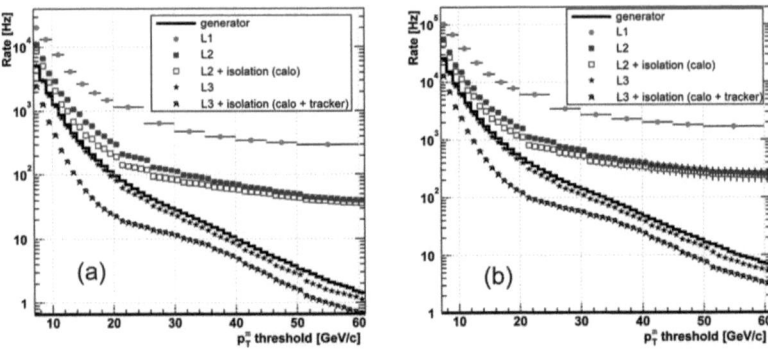

Figure 5.6.: The HLT single-muon trigger rates as a function of the p_T threshold for low luminosity ($2 \times 10^{33}\,\mathrm{cm}^{-2}\mathrm{s}^{-1}$, left) and high luminosity ($10^{34}\,\mathrm{cm}^{-2}\mathrm{s}^{-1}$, right) [34]. The rates are shown separately for Level 1 and various HLT levels. Track information is included at Level 3. The rate generated in the simulation is also shown.

The combinatorial background for muons is reduced in the High Level Trigger (**HLT**) by using track information from the tracking system as described in [35] and [34] and seen in figure 5.6. Associating tracker data already at an earlier stage of the trigger, should improve muon momentum measurement and the elimination of fake candidates. A hit in the tracker within a limited $\eta - \phi$ window should discriminate between ambiguous muon candidates and improve the p_T measurement.

For the electron trigger, the calorimeter isolation criteria alone is not sufficient for rejection. Again, the HLT uses additional information from the pixel system to select those electron candidates which did not originate from conversion or bremsstrahlung. It extrapolates energetic showers back to the pixel layers and tries to associate it with an isolated hit. Because of the high occupancy under sLHC conditions, it is unlikely to work well for the upgrade.

One major challenge for a tracking trigger is the impossibility to transfer all hits out of the detector within the small timeframe where the trigger decision has to be made. Data reduction on the detector level is therefore essential. On the other hand, the additional electronics needed to perform such a data selection, must not degrade the tracking performance by introducing to much additional material inside the tracker volume. This is very challenging, as a fast decision logic might be very power hungry which requires additional cables to supply the electric power and the additional power dissipation has to be removed from the tracker via cooling. It becomes clear, that it is imperative to minimize the material budget wherever possible, to make room for the required functionality that has to be added to the tracking system.

5.2. Challenges for the Strip Tracker

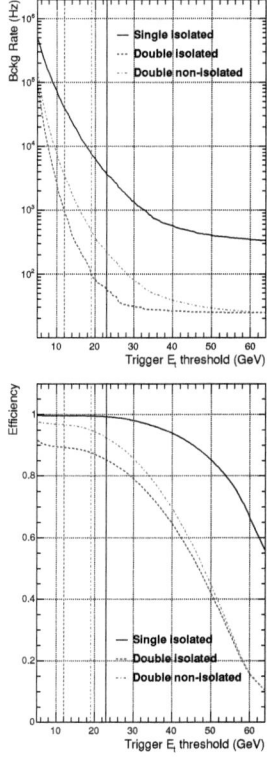
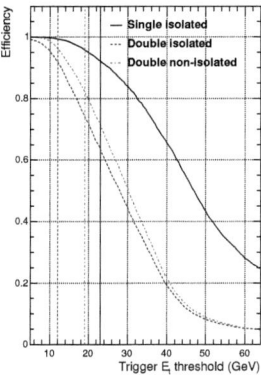

Figure 5.7.: The rates for electromagnetic triggers as a function of threshold for various types of event (left: jet events, middle: $H \Rightarrow 4e$, right $H \Rightarrow \gamma\gamma$) at a luminosity of $2 \times 10^{33} cm^{-2} s^{-1}$ [35]. Scaling to $10^{35} cm^{-2} s^{-1}$, it is clear, that typical single isolated electron rates would be far higher than tolerable for practical thresholds.

111

5.3. Summary

The challenges faced by the CMS Tracker under the proposed luminosity upgrade of the LHC accelerator, are mainly the higher track densities and the increase in irradiation. Both are caused by the increase in particle interactions in each collision which makes proper triggering much more difficult, ultimately requiring the tracker to provide information to the trigger system as well.

In this work I will review the current status on radiation hard silicon sensor and provide recommendations on which materials should be suitable for the middle and outer layers. I will describe the necessary tools which enable the testing of the process quality of such future radiation hard sensors. Higher granularity of the strip sensors is required to maintain a good pattern recognition performance. I will present one of the key concepts to enable high density strip sensors while reducing the material budget which is another important requirement for the upgrade of the CMS Tracker. I will not address solutions to the tracking trigger problem in this work, but research on this issue is currently starting [36] and the tools described in the following chapters will also be utilised to tackle the problem.

Radiation hardening is a method of designing and testing electronic components and systems to make them resistant to damage or malfunctions caused by high-energy subatomic particles and electromagnetic radiation, such as would be encountered in outer space, high-altitude flight, around nuclear reactors, or during warfare.

Wikipedia on **Radiation hardening**

6

New Bulk Materials and Process Technologies

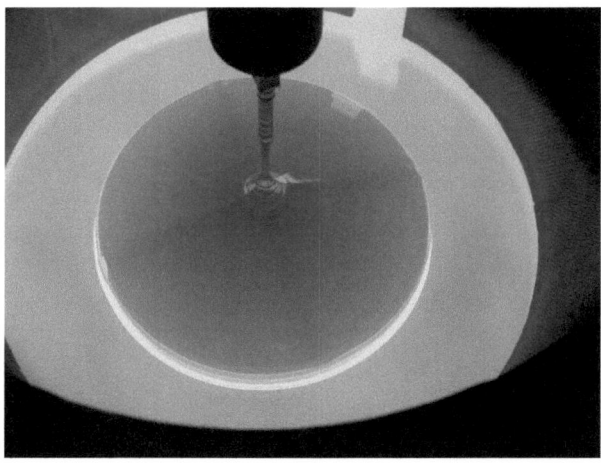

The significant increase in radiation due to the higher luminosity intended for the upgrade of the LHC (see section 5.2), will require new radiation hard bulk materials. Section 1.6 reviewed

6. New Bulk Materials and Process Technologies

the basic effects of radiation damage in silicon sensors, nevertheless many of the details are not yet fully understood.

CERN has established a dedicated R&D project in 2002 to search for appropriate materials entitled *"Radiation hard semiconductor devices for very high luminosity colliders"[37]* or RD50. We will review the most important results to select the most promising materials and process technologies suitable for the middle to outer regions of the tracker. The materials are expected to deliver enough electrons per MIP to ensure a minimum signal-to-noise ratio of 10 within the lifetime of the experiment (expected integrated luminosity: 2500 fb^{-1}).

6.1. Combined Results from RD50

Materials and processes which have been investigated and their typical properties are shown in table 6.1. I will review the results for different wafer materials and process technologies which were irradiated with protons or neutrons of various energies. The main focus will be the comparison of standard technology choices compared to their most promising new or advanced counterpart:

- Standard Float Zone (FZ) to Magnetic Czochralski (MCz) bulk material
- Standard 300 µm to thin 140 µm wafer thickness
- n-type to p-type bulk material
- p-type to n-type strip implants

High resistive float zone silicon has been the traditionally used bulk material for tracking detector applications despite its high cost and limited availability. Due to its high purity it allows operation at relatively low voltages due to its low full depletion voltage. However, in recent years it has been generally accepted, that bulk materials with enhanced concentration of oxygen shows an increased tolerance to gamma, electron and charged hadron irradiation [8].

It is possible to enhance the oxygen concentration in FZ material by diffusion from a SiO_2 coating of the wafer in a high temperature oxygen atmosphere (DOFZ material) [38], [39]. On the other hand, for silicon bulk material created by the Czochralski process, the intrinsic oxygen concentration is already much higher as seen in table 6.1 with the additional benefit of lower costs.

Many well established results are published, which demonstrate a suppressed introduction rate of acceptor-like defects in all oxygen enriched materials [40], [41]. Additionally, measurements on sensors irradiated with low energy reactor neutrons [42], high energy (23 GeV [43]) and low energy (10, 20 and 30 MeV [44]) protons and 190 MeV pions [41]), show some

6.1. Combined Results from RD50

Material	Type	Resistivity [kΩ]	V_{FD}	Oxygen [cm^{-3}]
Cz	n	< 0.001	> 30 kV (300 μm)	10^{18}
MCz	n	0.5 - 1	< 500 V (300 μm)	> 10^{17}
MCz	p	2	< 500 V (300 μm)	> 10^{17}
FZ	n	> 2	< 150 V (300 μm)	< 10^{16}
FZ	p	> 4	< 250 V (300 μm)	< 10^{16}
DOFZ	n	< 4	< 150 V (300 μm)	$1 - 2 \times 10^{17}$
Epi	n	< 0.25	<350 V (150 μm)	$\approx 10^{17}$
Epi	p	< 0.35	<900 V (150 μm)	$\approx 10^{17}$

Table 6.1.: Typical resistivities, full depletion voltage for the given detector thickness and oxygen concentration of different silicon crystals (Cz...Czochralski, MCz...magnetic Czochralski, FZ...float-zone, DOFZ...diffusion oxygenated float-zone, Epi...epitaxial layer).

clear advantages of MCz and Cz materials compared to FZ and DOFZ. Sensors made on epitaxial silicon grown on low resistive wafers show good performance after irradiation as first proposed in [45]. Nevertheless, some of these materials have disadvantages which make them unsuitable as tracking detectors. The Cz material usually has a full depletion voltage beyond practical values at a typical thickness of 300 μm, while the thin active area of epitaxial layers provides only a very small signal. Such devices may retain some usefulness in special applications, but for a strip tracking device like in CMS, a high signal-to-noise ratio at relatively low depletion voltages is very much favorable. Under such requirements, the most interesting material is magnetic Czochralski.

6.1.1. Full Depletion Voltage and Effective Doping Concentration

The full depletion voltage of a sensor is an important parameter to ensure proper operation of the device. As explained in chapter 2, the sensor bulk should be completely depleted of charge carriers to extract the maximum signal, while V_{FD} changes with irradiation. For materials going through type inversion, an operation voltage exceeding V_{FD} is even more essential, as they would cease to work as explained in section 1.6.2. Studies indicate, that for strip sensors the collected charge increases for voltages beyond V_{FD}[1] as reported in [46]. The maximum voltage that can be supplied to the sensors is limited by the HV-stability of the sensor and, even more so, by the capabilities of the power supply system in the actual detector. For the current CMS Tracker power supply system, which might have to be reused for the upgrade, this maximum voltage is 600 V as described in [47].

---
[1] The full depletion voltage V_{FD} is usually measured using the C-V method and is defined as the onset of the plateau.

6. New Bulk Materials and Process Technologies

Figure 6.1 shows the full depletion voltage V_{FD} of pad sensors as a function of neutron fluence. It is apparent, that the MCz materials exhibit a slower degradation in V_{FD} (≈ 55 V/10^{14} cm^{-2}) compared to standard FZ (≈ 125 V/10^{14} cm^{-2}).

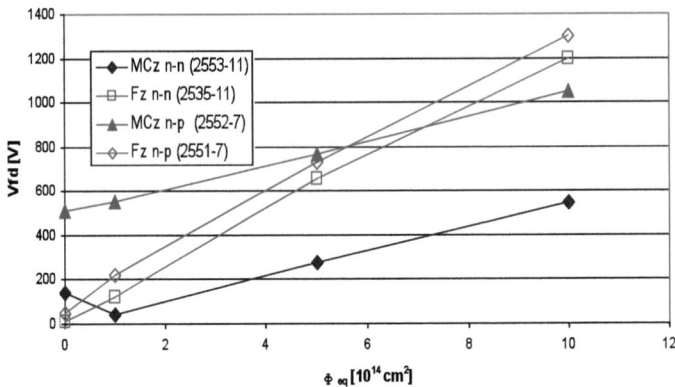

Figure 6.1.: V_{FD} as a function of neutron fluence for n- and p-type MCz and FZ pad sensors. Measurements performed using the C-V method. Neutron fluence has been normalized to 1 MeV neutrons. Pad sensors have been annealed for 80 min at 60°C [46].

The n-type MCz performs exceptionally well due to their higher initial resistivity and therefore lower initial V_{FD} and the additional benefit of donor removal until type inversion. After receiving a fluence of 10^{15} cm^{-2} the V_{FD} is approximately 500 V lower compared to p-type MCz while retaining the same slope.

Irradiation with protons shows slightly different results. As seen in figure 6.2, the change in $|N_{eff}|$ (V_{FD} respectively) of n-type MCz diodes, irradiated with 24 GeV protons does not show an advantage compared to p-type MCz. This is mainly because n-type MCz and Cz materials are not inverting under proton irradiation, while p-type MCz and n-type DOFZ do and therefore tend to have a lower V_{FD}. The changes of N_{eff} in the non inverting materials suggests, that the defects introduced by proton radiation are mainly donors in contrast to the acceptor-like defects introduced in FZ materials.

Figure 6.3 shows the changes of N_{eff} as a function of proton and neutron fluence for different materials and different thicknesses of detectors. It can be seen that the n-type MCz 100 μm thick diodes do not invert after proton irradiation and invert after neutron irradiation, behaving like thick diodes. More surprising are the results of the FZ devices: the 100 μm thick ones show inversion with both neutron and proton irradiations, as expected, while the 50 μm thick

6.1. Combined Results from RD50

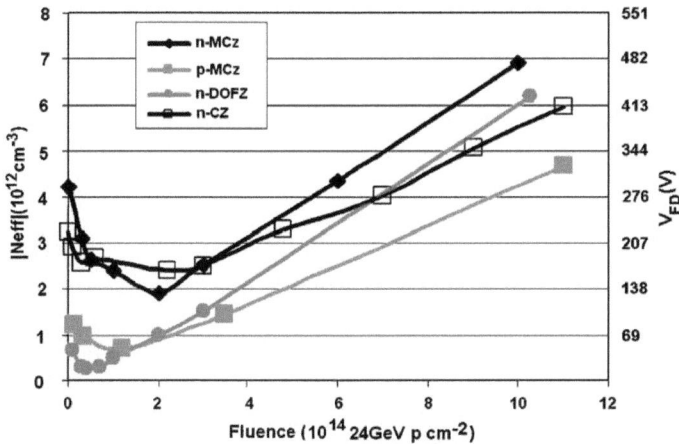

Figure 6.2.: Absolute value of N_{eff} (V_{FD} respectively) as a function of proton fluence for n- and p-type MCz and n-type oxygenated FZ and CZ sensors. Measurements performed with the C-V method [48].

n-type FZ devices do not show inversion after proton irradiation. This effect of the thickness on the FZ material is not yet understood.

An interesting observation can be made for the n-type MCz devices: from figure 6.3a, it appears that detectors being irradiated with both neutron and proton are building-up opposite charged damage defects. The defects can partially compensate each other in the case of proton and neutron simultaneous irradiation, resulting in a significantly lower degradation rate of V_{FD}. This is actually a real condition in the sLHC, where the radiation flux is composed of charged particles emerging from the interaction region and by backscattered neutrons from the calorimeter volume. These fluxes are equal at about 25 cm radius from the interaction point inside the experiment. Detectors located at about this distance from the interaction point could benefit from the cancellation effect of the opposite signed defects if equipped with the n-type MCz sensors.

The different behavior under neutron and proton irradiation as described in the previous paragraphs shows, that the NIEL hypothesis (see section 1.6) is not applicable anymore to new materials under heavy irradiation!

6. New Bulk Materials and Process Technologies

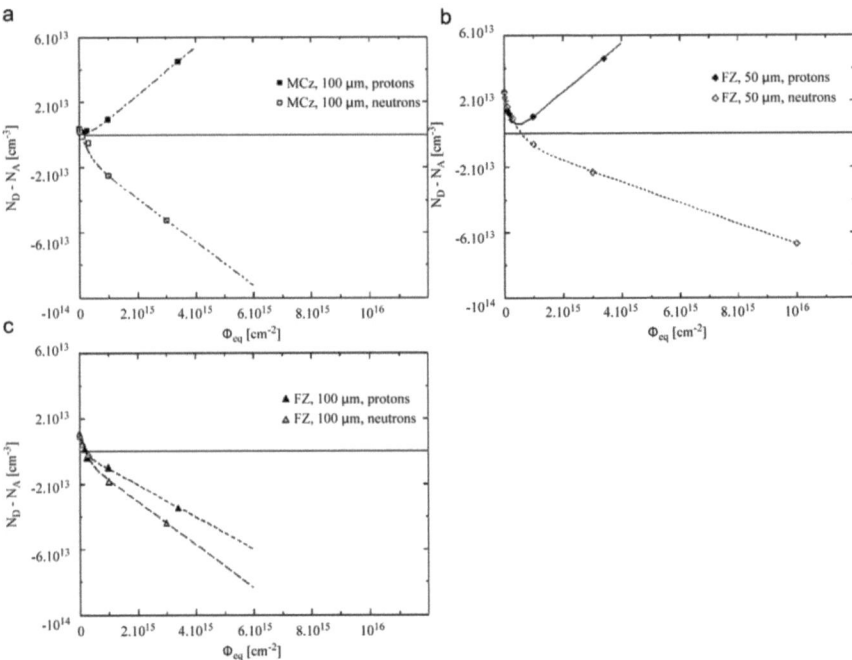

Figure 6.3.: Changes of N_{eff} as a function of n and p irradiations for different diodes [49]: 100 μm n-MCZ (a), 50 μm n-FZ (b) and 100 μm n-FZ (c).

6.1.2. Charge Collection Efficiency

One of the crucial parameters to judge the capabilities of a sensor, is the amount of collected charge per incident particle or Charge Collection Efficiency (**CCE**). Depending on the readout chips, a certain minimum charge is needed to achieve a good signal-to-noise ratio. As the CCE degrades with irradiation, it is one of the factors defining the lifetime expectancy of a strip sensor in a high radiation environment.

From figure 6.4 it becomes evident, that the extracted charge from strip sensors increases with bias voltages beyond V_{FD} as determined in figure 6.1. Nevertheless, the absolute voltages needed at irradiations doses around 10^{15} 1 MeV neutrons per cm^2 become impractical. A solution to produce sensors which require a low V_{FD} despite having a low resistivity due to irradiation damage, is the usage of thin substrates for the sensors (see section 1.5.1).

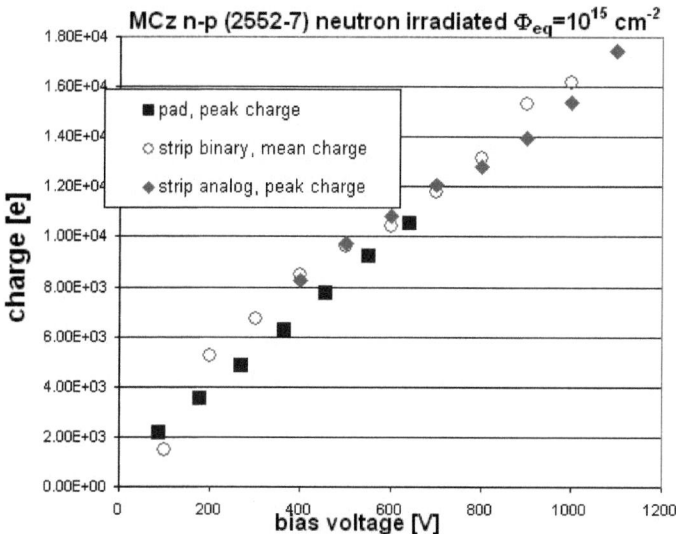

Figure 6.4.: Mean collected charge as a function of bias voltage for MCz pad and strip sensors. Neutron fluence for all structures was 10^{15} 1 MeV neutrons per cm^2 and annealing for 80 min at 60°C[46]

Thin (140 μm) and thick (300 μm) strip sensors have been irradiated up to 10^{16} 1 MeV neutrons per cm^2 [50]. As seen in figure 6.5 the charge collected for thick devices at high bias voltages exceeds the performance of the thin ones. This suggests that the charge collection distance is longer than 140 μm even at such high doses.

6. New Bulk Materials and Process Technologies

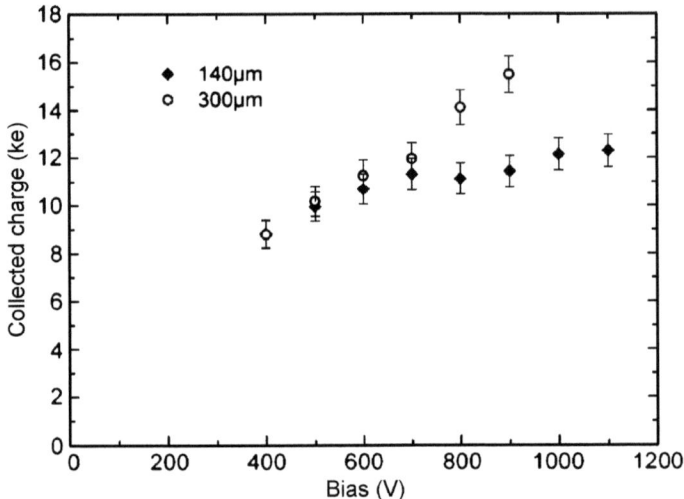

Figure 6.5.: Collected charge as a function of the bias voltage for thin (140 μm) and thick (300 μm) detectors irradiated to 1.6×10^{15} 1 MeV neutrons per cm^2 [50].

At the very high dose of 10^{16} 1 MeV neutrons per cm^2 as seen in figure 6.6 the pictures changes and favors the thin sensors. This suggests, that the charge collection distance has fallen below 140 μm. One should keep in mind, that thin detectors irradiated up to 10^{16} 1 MeV neutrons per cm^2 can still provide a signal exceeding 7.000 electrons at bias voltages above 600 V as also seen in figure 6.6

6.1.3. Reverse Bias Current

Another limiting factor determining the performance and lifetime of strip sensors, is the amount of reverse bias current. It is strongly dependent on the bulk damage caused by irradiation as explained in section 1.6. With more current drawn by the sensors, the ohmic losses in the cables increase and require more cooling. More importantly the higher current will increase the temperature of the silicon, in turn causing an increase of the reverse bias current. This vicious circle called *thermal runaway*, will render the detector unusable and therefore limit their lifetime.

Another advantage of thin sensors, is the reduced reverse bias current. This is due to the smaller active bulk volume available to generate the current compared to thick sensor. The

6.1. Combined Results from RD50

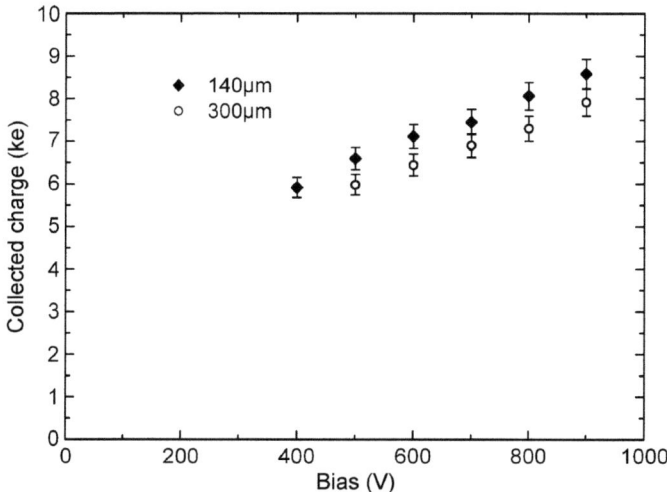

Figure 6.6.: Collected charge as a function of the bias voltage for thin (140 μm) and thick (300 μm) detectors irradiated to 10^{16} 1 MeV neutrons per cm^2 [50].

plots in figure 6.7 clearly show the expected behavior for highly irradiated strip sensors at high bias voltages.

6.1.4. Charge Multiplication in Silicon

Recently a number of measurements hinted on an effect which was previously unknown in silicon – charge multiplication. A seen in figure 6.8 the signal extracted from thin epi-sensors at high reverse bias voltages is significantly larger than for a fully depleted device after low irradiation.

Even after closer investigation as presented in [51], the charge multiplication effect seems to be true. Nevertheless, it is not yet clear if it can be used to improve the signal-to-noise ratio in an actual detector, as the impact on noise is not yet clear. Furthermore, the charge multiplication effect could be accompanied by other detrimental effect like micro-discharges (as reported for example in [52]), making the exploit of the charge multiplication unfeasible.

6. New Bulk Materials and Process Technologies

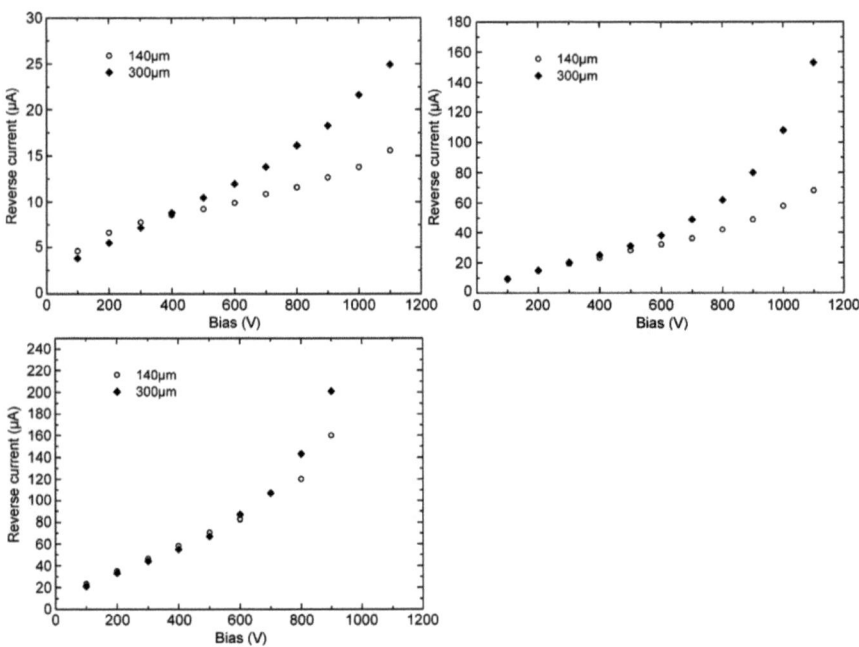

Figure 6.7.: Reverse current as a function of the bias voltage for pairs of 300 μm and 140 μm thick sensors irradiated to 1.6 (top), 3 (middle) and 10 (bottom) $\times 10^{15}$ 1 MeV neutrons per cm^2 [50].

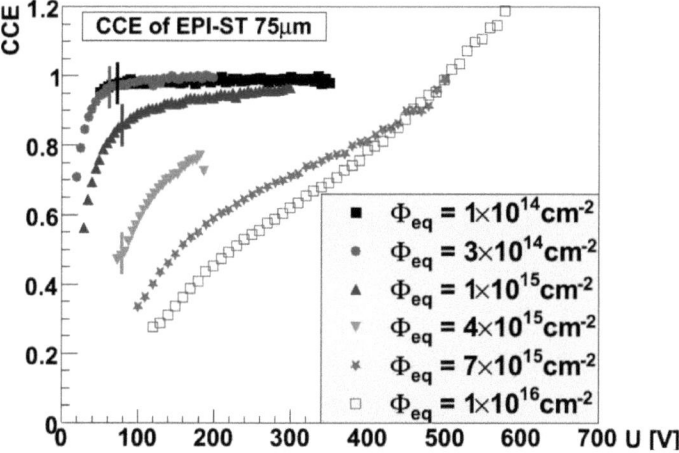

Figure 6.8.: Charge collection efficiency for thin epi-sensor after various irradiation doses. The CCE for highly irradiated sensor can exceed one at high reverse bias voltage hinting on a charge multiplication process [51].

6.2. Summary

Standard sensor with p+ strip readout in n-type FZ material are not suitable for the high irradiation doses received by the inner layers of the tracker in an sLHC environment. They should survive in the outer layers of the tracker in the sLHC environment, where they will receive a dose similar to the inner layers of the current tracker integrated over the lifetime of the LHC ($\approx 10^{14}$ 1 MeV neutrons per cm^2). Previous works have shown, that full detector modules irradiated up to 0.65×10^{15} 1 MeV neutrons per cm^2 will still provide a signal-to-noise ratio well above 12 at medium depletion voltages of 400 V [53]. Experience with the current tracker will give more evidence on the performance of standard p-on-n FZ sensors.

Nevertheless, the middle to inner layers of the tracker will receive significantly higher doses in the sLHC environment. Alternative materials and processes such as MCz, thin bulk materials or n strip readout offer promising properties to withstand such high irradiation doses. They show improvements in the key aspects defining the lifetime and performance of a detector: effective doping concentration (full depletion voltage), charge collection efficiency and reverse bias voltage.

From the results presented above, thin n-type or p-type MCz with n+ readout strips show the most promising results. Nevertheless, these advantages have to be balanced with the real

6. New Bulk Materials and Process Technologies

production costs of such sensors. The non-standard material and manufacturing process will have an impact on the manufacturing costs. Additionally, only a limited number of vendors is currently able to produce such sensors – if only in small quantities, let alone the large number of sensors needed to equip the CMS tracker volume.

For the outer layers, where the amount of irradiation is more relaxed but the area needed to be equipped with sensors is larger, standard p+ strip readout on n-type FZ or MCz might prove a more cost effective solution without compromising performance.

6.3. Outlook

To conclude the selection of a suitable material and manufacturing process, a production of several test sensors and structures at Hamamatsu, Japan was devised [54]. The same structures will be manufactured on all currently available technologies mentioned in this section.

The measurements of these test sensors and structures during an extensive irradiation campaign, will give us the capability to finally select the most adequate technology for the different areas of the tracker. With Hamamatsu, one of the major suppliers capable of providing high quality sensor on a mass production scale was contracted to perform this order. It will also help the company to provide us with a better understanding of the costs involved with each of the technologies.

> *Quality assurance, or QA for short, refers to a program for the systematic monitoring and evaluation of the various aspects of a project, service, or facility to ensure that standards of quality are being met.*
>
> Wikipedia on **Quality Assurance**

7
Quality Assurance using Test Structures

An important task during the production of silicon strip sensors, be it prototypes for test purposes or the mass production to equip a large detector like the CMS Tracker, is the implementation of a comprehensive quality assurance procedure. During the mass production of the sensors for the current CMS Tracker, the collaboration used a set of purpose-built standard test structures called *halfmoon*. They were included on the cut-aways towards the edge of each wafer and proved to be an indispensable tool [28].

The results gained from electric measurements on test structures enable a deeper insight into the performance of irradiated materials as well. During irradiation campaigns, where sensors and test structures are irradiated with neutrons and protons to several different doses, the test

7. Quality Assurance using Test Structures

structures provide easy access to the most important parameters of a sensor and complement the measurements done on the actual sensors. They enable a closer diagnostic of the damage done to the sensor material.

The design of the test structures for the production of the current CMS Tracker had certain shortcomings which made some measurements unreliable. Furthermore, certain parameters were not accessible on the original halfmoon, making the design of additional structures desirable. Finally, the standard set of test structures was only available as a ready-made design implemented for the process used to manufacture the sensors for the current CMS Tracker and they could not be used on most of the new materials the collaboration is interested today.

7.1. Revising and Enhancing the Standard Set of Test Structures

The absolute values of some parameters measured on the structures are very small, and are challenging to measure even with well designed and calibrated equipment. Additionally, some structures were never used and became obsolete.

The original design and test procedures are described in [28]. The changes are documented for each structure in the following sections.

Sheet

The *sheet* structure provides access to the resistivity of the materials implanted or deposited on the wafer. For very low resistive materials such as deposited aluminium, the absolute resistances become very small, in the order of the contact resistance of the needles to the aluminium pad. Two improvements were implemented to provide a more reliable measurement:

7.1. Revising and Enhancing the Standard Set of Test Structures

- The length of the aluminium meanders was increased to raise the absolute resistance to be measured.
- The size of the contact pads was increase significantly to provide enough space for two needles on the same pad. This enables the use of a four-wire measurement to compensate the contributions from the wiring and the contact resistance.

Furthermore, the new design can be configured to allow access to the resistivity of all relevant materials which are created or deposited during the manufacturing process. Additional meander-like structures can be included within the structure where the exact selection of materials to be measured depends on the specific process.

GCD

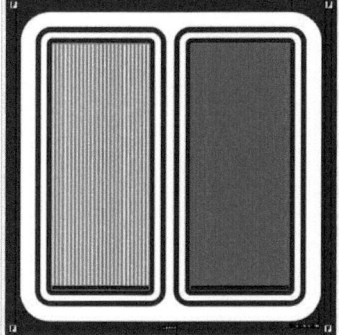

The original *GCD* structure contained four gate controlled diodes of which the round ones were never measured, as they offered no additional information over the square shaped GCDs. The free space was used to enlarge the remaining rectangular diodes which also improved the reliability of the measurement. This is due to the larger surface area of the structures and the therefore increased dark current through the diode.

Cap_TS_AC

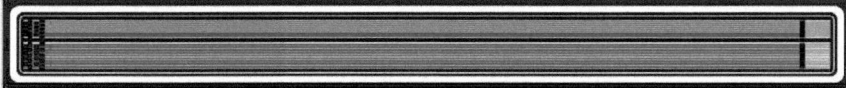

The *Cap_TS_AC* structures is used to measure the capacitance between a strip and its six nearest neighbours. The absolute value is very small for the short strips used in the original

7. Quality Assurance using Test Structures

version, at approximately 1 pF. It is a very challenging task to reliably measure such small capacitances. By elongating the strips, the interstrip capacitance could be increased. The structure was rotated by 90 degrees to use the available space on a typical halfmoon-shaped cutaway of a wafer more efficiently. The length of the Cap_TS_AC should correspond to the combined width of all the other structures.

Diode

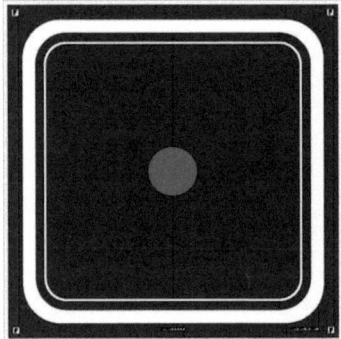

A standard diode for the measurement of the full depletion voltage (and the bulk resistivity respectively) is an essential tool in any set of test structures. Unfortunately, the original diode in the CMS test structures had a poor design, preventing operation at high voltages. Irradiated diodes exhibit an increase in full depletion voltage as explained in section 1.6. The original diodes were not usable after high irradiation as they experienced breakdown before being fully depleted. The new design should make the structure much more high voltage robust by avoiding sharp corners, modifying the periphery of the structure and adding metal overhangs similar to the breakdown protection strategies described in section 2.2.4. This should enable the usage of the diode also for irradiation campaigns to record IV-curves up to 1000 V.

MOS

In the original halfmoon two MOS structures were included. One should measure the quality of the thin readout (or gate) oxide while the second one should investigate the thick interstrip (or field) oxide. For the process used to manufacture the current sensors of the CMS Tracker, it is not possible to produce a MOS structure with the thin readout oxide. Therefore the CMS Halfmoon contained two identical MOS structures with thick oxide. So far all manufacturing processes have the same limitation making a single MOS structure sufficient.

New structure: CapDM

A new structure was designed which is only feasible for processes with two top metal layers. *CapDM* is then used to measure the thickness of the oxide between the metal layers. It is basically a parallel plate capacitor, with large rectangular electrodes on each of the metal layers which can be accessed using contact pads as illustrated in figure 7.1

7. Quality Assurance using Test Structures

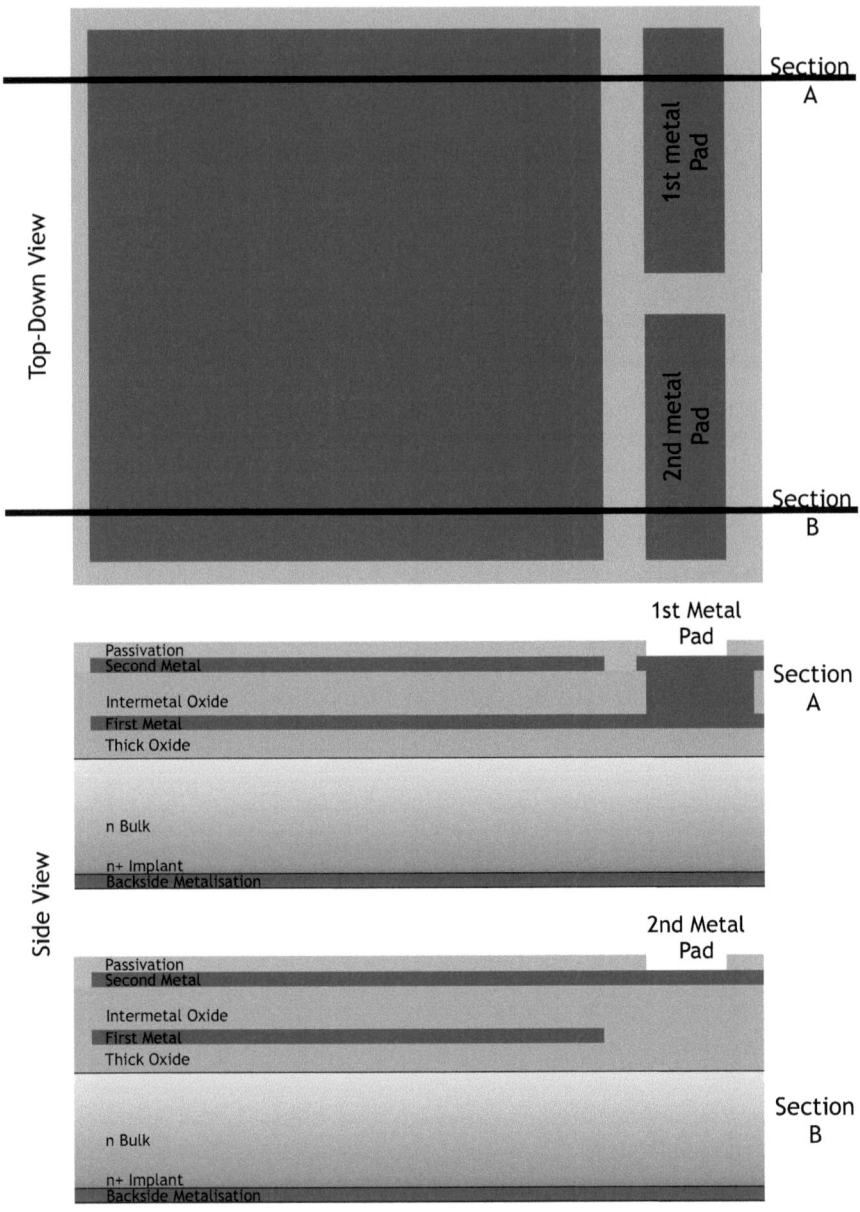

Figure 7.1.: Illustration of the CapDM structure from a top-down view (upper) and side views in two different sections (lower). The layer thicknesses in the side views are not to scale!

7.2. First Production of the Enhanced Set of Test Structures

In 2007 a very first run of three wafers, each with five sets of test structures was produced at ITE Warsaw[1] as seen in figure 7.2. Only part of the above mentioned improvements were implemented and this production was only considered to be a starting point for further collaboration. The improvements mentioned in the previous section were implemented for the diode, the sheet structure and the GCD, where the design of the diode was adopted from RD50. Figure 7.3 shows a screenshot from the original mask design.

Figure 7.2.: Photo of one of the wafers delivered by ITE Warsaw. Each wafer contained 5 full sets of test structures.

Two goals that were targeted with this production:

- Verify the improvements made on the test structures themselves.

---
[1]**Instytut Technologii Elektronowej**, Al. Lotnikow 32/46, 02-668 Warszawa

7. Quality Assurance using Test Structures

- Validate the manufacturing process implemented by ITE Warsaw.

This production was the first collaboration between ITE Warsaw and the HEPHY Vienna. A continuing collaboration between the two institutes was envisaged and this production served as a starting point to produce full featured strip sensors at ITE Warsaw for R&D prototypes. Therefore, the validation of the manufacturing process was the main target, while the improvements in the test structures where only secondary. As the effects originating from the design of the structures themselves and those caused by peculiarities in the manufacturing need to be disentangled, a conservative approach was taken by implementing only few of the improvements suggested above.

Figure 7.3.: Picture of the masks of the full set of test structures, extracted from the original mask design as prepared for the first run with ITE Warsaw. From left to right: **TS-Cap, Sheet, GCD, Cap-TS-AC, Cap-TS-DC, Diode, MOS1, MOS2**

7.2.1. Results

The results of the electric measurements on the test structures were first published in [55].

Diode

The diode was derived from the design used in the RD50 collaboration [37] featuring multiple guard rings and rounded corners. This layout showed high voltage robustness in many other production runs (see any RD50 publication on *diodes* or *pad detectors* like [46]). In figure 7.4 the diodes measured on five sets of a single wafer show a very robust current-voltage curve up to and beyond 600 V. The outlier in set A does not detriment this conclusion, as a design flaw should be evident in all structures. High electric fields caused by a poor layout, would set an upper voltage limit, at which breakthrough occurs, regardless of the production quality. As most of the diodes reached much higher voltages than the old design, which was only capable

7.2. First Production of the Enhanced Set of Test Structures

of reaching a maximum of ≈ 350 V before breakdown, the implemented improvements were successful.

Nevertheless, the quality of the manufacturing process limited high voltage robustness as diodes of the same design but different manufacturer were operated successfully at much higher voltages.

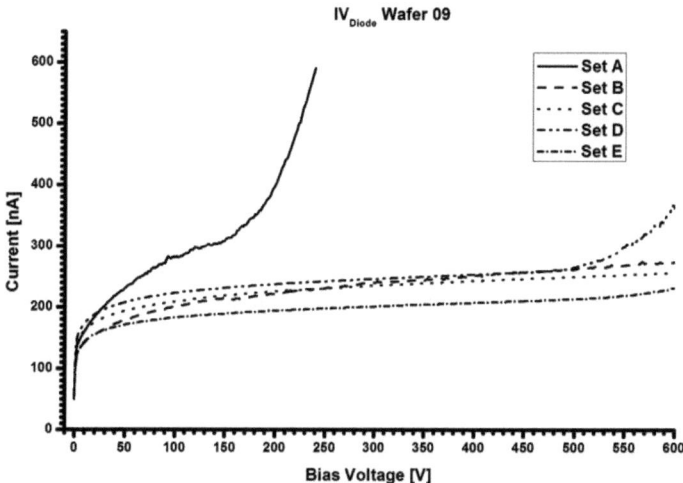

Figure 7.4.: Current-Voltage curve for the five diodes of a single wafer. Most of the diodes a very stable up to 600 V and beyond. One structure suffers from early breakdown which might be attributed to defects caused by the manufacturing process, but not to the layout of the diode itself.

Sheet

The larger contact pads in the sheet structure offered sufficient space for two needles, making a four-wire-measurement feasible. Additionally, the length of the aluminium strips was increased and the absolute resistance increased to ≈ 250 Ω (20 μm width) and ≈ 1 kΩ (10 μm width). The resistivity of the aluminium deposited on the wafer is depending on the process parameters during production, again influencing the final value of the resistance and its distribution on the wafer. Nevertheless, the electric measurement itself was considerably more reliable, meaning that repeated measurements on the very same strip gave the same result with only small deviations. For completeness, the measured resistivities are shown in table 7.1.

7. Quality Assurance using Test Structures

Wafer/Set	Resistivity		Wafer/Set	Resistivity		Wafer/Set	Resistivity	
	20 µm	10 µm		20 µm	10 µm		20 µm	10 µm
02/A	58	70	04/A	55	65	09/A	51	57
02/B	59	71	04/B	54	62	09/B	50	57
02/C	58	71	04/C	53	60	09/C	48	56
02/D	50	58	04/D	52	60	09/D	48	53
02/E	48	55	04/E	50		09/E	49	55
Average	54,6	65	Average	52,8	61,8	Average	49,2	55,5
Overall Average	52,2	60,7						

Table 7.1.: Resistivity values measured on all aluminium sheet structures. Each structure contained two strips with different widths (10 and 20 µm respectively). Due to a defect in the structure in set E on wafer 04, no value could be extracted.

GCD

The GCD structure had already received the modifications explained in the preceding section. The larger area resulted in a higher reverse bias current making the measurement of the extracted signal more reliable. A typical measurement is shown in figure 7.5.

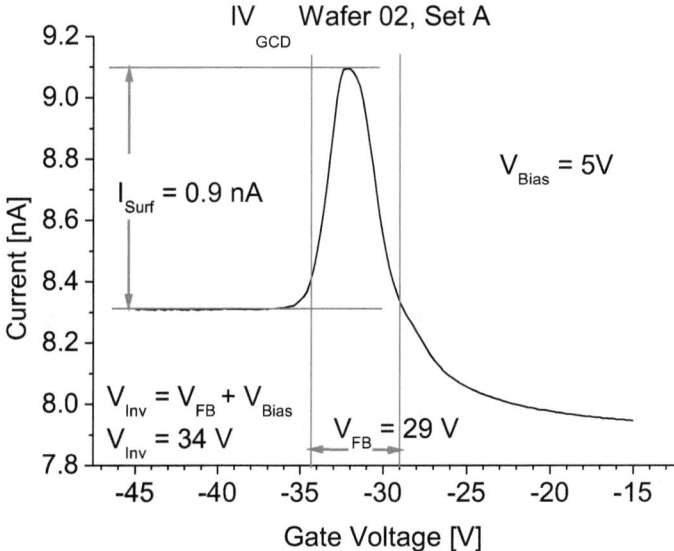

Figure 7.5.: Current through the diode as function of the gate voltage of the GCD. Flatband voltage and surface current are the signals which are extracted and are shown in the plot.

7.3. Second Production of the Enhanced Set of Test Structures

Following the first production, the implementation of all foreseen enhancements of the structures was finished beginning of 2009. At that stage, the full set of test structures was already requested by several collaborators of the institute and other third parties to be included on their production runs. The delivery, measurement and analysis of these wafers will happen after the finalisation of this work and are therefore not in the scope of this thesis.

A new production was also started with ITE Warsaw (ITE09). As mentioned earlier, the second run concentrated on strip sensors and is explained in more detail in chapter 8. Nevertheless, the complete set of enhanced test structures was placed on the wafers, including the CapDM structure as seen in figure 7.6. This gave us the opportunity to evaluate the full set of enhanced test structures which is described in this section. Additionally, they provided the necessary measurements to judge the manufacturing process which will be discussed in section 8.4.1.

The run was split into two different process implementations: one half of the wafers were

7. Quality Assurance using Test Structures

Figure 7.6.: Picture of the masks of the full set of test structures, extracted from the original mask design as prepared for the second run with ITE Warsaw (bottom) and a photo from the final structures cut from the wafer (top).. From left to right: **CapDM, TS-Cap, Sheet, GCD, Cap-TS-DC, Diode, MOS** and at the bottom **Cap-TS-AC**.

processed in a double metal process, while the other half was stopped after the first metal layer and only the passivation was added just like in a standard single metal process. This gave us the opportunity to have the test structures manufactured in a standard single metal process and in a double metal process where most of the production steps were identical between the two. This way, the modifications necessary to make the set of test structures work in a double metal process could be validated as well.

7.3.1. Results

The test structures from this second production run with ITE Warsaw went through the same characterisation as the previous one. The main target was the performance of the modified Cap-TS-AC and Diode structure and the entirely new CapDM structure. Some additional measurements were made to asses the production quality and point out any severe deviations from the expected values.

Diode

The new diode was designed to withstand a reverse bias voltage up to 1000 V, which would enable the usage of this structure in high irradiation campaigns. As seen in figure 7.7 the structure is clearly able to achieve a low dark current even beyond that requirement up to the maximum of 1100 V that our laboratory power supplies can provide. The quality of the manufacturing process seems to limit the voltage robustness earlier than the design would allow, as the instabilities at the top end of the plots in figure 7.7 are disappearing if the structures are given enough time to settle between measurements. Some of the otherwise identical diodes on the different wafers are not stable beyond a few hundreds of volts, which is especially true for the wafers going through the second metallisation process. This indicates instabilities in the manufacturing process, nevertheless the design of the diode is robust towards high depletion voltages for the single and double metal implementation

The measurement of a $\underline{C}$apacitance-$\underline{V}$oltage (CV) curve on the diode reveals the full depletion voltage V_{FD}. In figure 7.8 the curves reveal the same shape on all wafers with the exception of wafer 4 which exhibits a strange rise in capacity after full depletion. Characteristic to all curves are two knees which hints on a non-isotropic doping profile of the wafer. The full depletion voltage extracted from the curves is around 24 V.

7. Quality Assurance using Test Structures

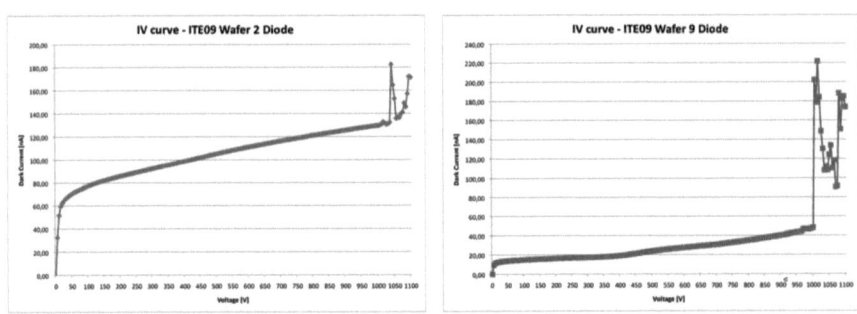

Figure 7.7.: Exemplary IV curves of two diodes from single metal wafer 2 (left) and double metal wafer 9 (right). Both structures are stable up to and beyond 1000 V.

Cap_TS_AC

The improved design of the Cap_TS_AC structures provides significantly longer strips which should increase the measured signal, the interstrip capacitancy C_{int}. The strip length in the original CMS Halfmoon was 10.6 mm while the new version has a strip length of 35 mm, an elongation of a factor of 3.3. Table 7.2 shows the measured interstrip capacitances for a structure from the CMS production against the values from the wafer 3 of the ITE09 production. The measurements yield a factor of 3.2 to 3.7 difference in C_{int} which, taking the slightly different pitches and the different production processes into account, agrees well with the exceptions. This increase in absolute capacitance improved the reliability of the measurement significantly, which was the goal of modification to the design.

Structure	Capacity		Factor C_{ITE}/C_{CMS}	
	75 kHz	1 MHz	75 kHz	1 MHz
CMS 83 µm pitch	1.39	1.42	3.4	3.6
ITE 80 µm pitch	4.66	5.14		
CMS 122 µm pitch	1.66	1.68	3.2	3.7
ITE 120 µm pitch	5.32	6.14		

Table 7.2.: Measured interstrip capacitances (C_{int}) on the Cap_TS_AC structure of a standard CMS Halfmoon from the original CMS sensor production compared to the improved version of the structure from the ITE09 production at ITE Warsaw (wafer 3). Both structures provide two different pitches and they have been measured with a precision LCR meter (Agilent HP4285A) at a frequency of 75 kHz and 1MHz.

The wafers processed with two metal layers show a slightly higher interstrip capacitance as seen in table 7.3. The source of this increase in capacitance can only be caused by the

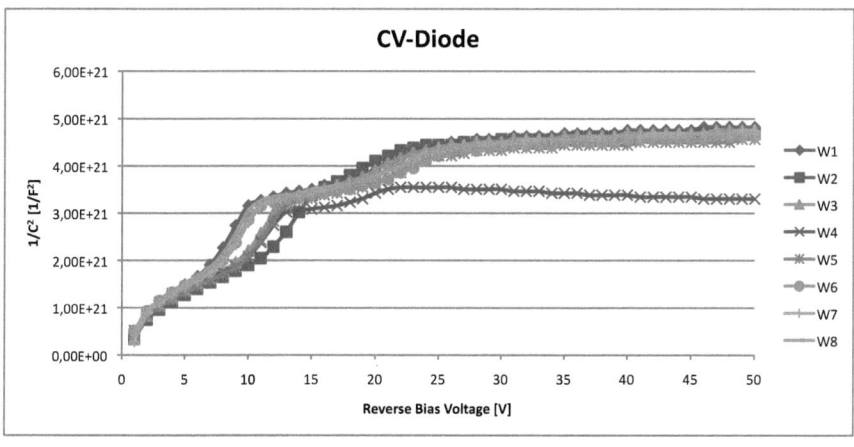

Figure 7.8.: CV curves for a single diode on each wafer. The capacity has been measured using an LCR meter at 1 kHz.

additional metal layer in the large contact pads.

wafer	narrow pitch (80 μm)		wide pitch (120 μm)	
	75 kHz	1 MHz	75 kHz	1 MHz
1	5.58	6.02	4.83	5.27
2	5.17	5.6	4.68	5.15
3	5.32	6.14	4.66	5.14
6	5.93	6.44	5.37	5.88
7	5.77	6.42	5.17	5.83
8	5.81	6.21	5.14	5.66

Table 7.3.: Measured interstrip capacitances (C_{int}) on the improved Cap_TS_AC structure from the ITE09 production at ITE Warsaw. The structure provides two different pitches and they have been measured with a precision LCR meter (Agilent HP4285A) at a frequency of 75 kHz and 1MHz. Wafers 1 to 3 are produced using a standard single metal process, while wafers 6 to 8 feature a second metal layer.

CapDM

The CapDM structure is a new design with the task to determine the thickness of the oxide between the two metal layers. It should provide a reliable measurement of the capacitance between electrodes in the first and second metal layer which are separated by a thin layer of dielectric. At given area of the electrodes, the capacitance is then defined by the permittivity

7. Quality Assurance using Test Structures

and thickness of the oxide layer:

$$C = \frac{\varepsilon_0 \cdot \varepsilon_{ox} \cdot A}{d}, \qquad (7.1)$$

where $\varepsilon_0 = 8.85 \times 10^{-12}$ F/m, $\varepsilon_{ox} = 3.9$ for SiO_2 and the area of the electrodes is 5×5 mm^2 = 25×10^{-6} mm^2.

In table 7.4 the results from the measurements of the CapDM structures are shown. The corresponding oxide thickness calculated from the average capacitance yields 600 nm which corresponds very well to the 700 nm before reflow as specified for the manufacturing process in table 8.1.

wafer 5	wafer 6	wafer 7	wafer 8	Average
1.422 nF	1.437 nF	1447 nF	1.384 nF	1.422 nF
			average oxide thickness	607 nm

Table 7.4.: Results from the capacitance measurements of the CapDM structure using a LCR meter at 1 kHz. The structure is only fully implemented on wafers with two metal layers and each wafer hosts four identical copies. The average oxide thickness is calculated using equation 7.1.

TSCap

The TSCap structure enables the measurement of the capacitance between implant and readout metal. The results are shown in table 7.5. The thickness of the oxide layer in-between can be derived from the result using equation 7.1 again. It is clearly evident from the data, that the double metal process yields a slightly thicker oxide than the single metal one, although both process are exactly the same up to the first metal layer. The additional process steps to apply the second metal layer seem to influence the underlying oxide layers.

More important is the absolute thickness of the oxide layer which is much thicker than expected for the thin coupling capacitance. Usually, a thickness of around 100 - 200 nm is expected, while the measurements reveal a much thicker oxide of around 700 nm.

Sheet

The sheet structure gives access to the resistivities of all deposited or doped materials on the wafer. The successful improvements to the structure were already demonstrated in the preceding section. Nevertheless, the resistivity of the polysilicon was measured, as it pointed out an interesting flaw in the manufacturing process.

As shown in table 7.6 the resistivity of polysilicon deposited on the wafer is very low. The expected value was around 6 to 7 kΩ/sq which is orders of magnitude higher than the measured

single metal	wafer 1	wafer 2	wafer 3	wafer 4	Average
	7.9 pF	8.0 pF	7.8 pF	8.0 pF	7.9
			average oxide thickness		653 nm
double metal	wafer 5	wafer 6	wafer 7	wafer 8	Average
	7.3 pF	7.3 pF	7.5 pF	7.3 pF	7.35
			average oxide thickness		702 nm

Table 7.5.: Results from the capacitance measurements of the TSCap structure using a LCR meter at 1 kHz and a reverse bias voltage of 30 V. The capacitance and therefore the derived oxide thickness is different for single and double metal processes. The average oxide thickness is calculated using equation 7.1.

resistivity.

single metal	wafer 1	wafer 2	wafer 3	wafer 4	Average
	207 kΩ	147 kΩ	137 kΩ	160 kΩ	163 kΩ
			average polysilicon resistivity		0.1 kΩ/sq
double metal	wafer 5	wafer 6	wafer 7	wafer 8	Average
	163 kΩ	185 kΩ	172 kΩ	146 kΩ	166.5 kΩ
			average polysilicon resistivity		0.1 kΩ/sq

Table 7.6.: Results from the resistivity measurements of polysilicon in the sheet structure. The resistance was measured on each wafer and the average resisitivity was then calculated for wafers with single and double metal processing individually.

7.4. Summary and Outlook

The usage of test structures to asses the quality of the manufacturing process, wether during the commissioning of a new vendor, to perform measurements in an irradiation campaign or as QA monitoring during the mass production, has proven to be a valuable tool. The structures where enhanced to improve the reliability of the measurements successfully while new structures have been designed and implemented to measure additional parameters.

The set of test structures was already produced and tested in two production runs at ITE Warsaw. The improvements suggested and implemented by me where tested successfully while the results of the electrical characterisations yielded valuable information on the quality of the production process. These conclusions will be discusses in the section 8.4.1.

Additional new structures will be designed to measure parameters like via resistances or strip capacitances introduced by second metal layer routings. Furthermore, the full set of test structures is currently being optimized for manufacturing processes like n-on-n, n-on-p and

7. Quality Assurance using Test Structures

double-sided processing. Major modifications are necessary to allow access to the same electrical parameters in such fundamentally different processes. Again, the SiDDaTA framework is flexible enough to allow these modification to be made.

Furthermore, the current set of test structures is used by several collaborations and is in production at different manufacturers. An important usage will be the production on different thin wafer materials and different processes at Hamamatsu, Japan [54]. Es mentioned earlier, the electrical characterisation of irradiated test structures will play an important role in selecting the most appropriate material and manufacturing process for sensors for the CMS Tracker upgrade.

Routing is a crucial step in the design of integrated circuits. It builds on a preceding step, called placement, which determines the location of each active element of an IC. After placement, the routing step adds wires needed to properly connect the placed components while obeying all design rules for the IC.

Wikipedia on **Routing (electronic design automation)**

8

Sensors with Integrated Pitch Adapters

The pitch of the strips on a sensor is usually different from the pitch of the connection pads on the readout chip. In CMS , the pads on the APV25 chip are in two staggered rows with a pitch of 44 μm, while the pitch of the sensor strips varies from 80 to 200 μm. This makes the use

8. Sensors with Integrated Pitch Adapters

of an additional adapter necessary as illustrated in figure 8.1. In traditional detector module designs, this P̲itch A̲dapter (**PA**) is made of thin aluminium lines deposited on a glass substrate as seen in figure 8.2. Both ends of the PA have to be wire bonded to the chip and the sensor respectively. See section 4.3 for the current module design in CMS.

The PA adds to to the overall material budget and it doubles the amount of bonds necessary to connect the electronics to sensor. Furthermore, it limits the flexibility of the module design, as the chips and the PA can only be located at the edge of the sensor. The connection of sensors with a very high number of strips, for example by segmenting each strip into several shorter ones, becomes difficult. At the maximum, all four edges of the sensor could be equipped with PAs and readout chips, but the integration of such modules in the detector would be very difficult. Therefore a new strategy for connecting readout chips to the sensors is essential.

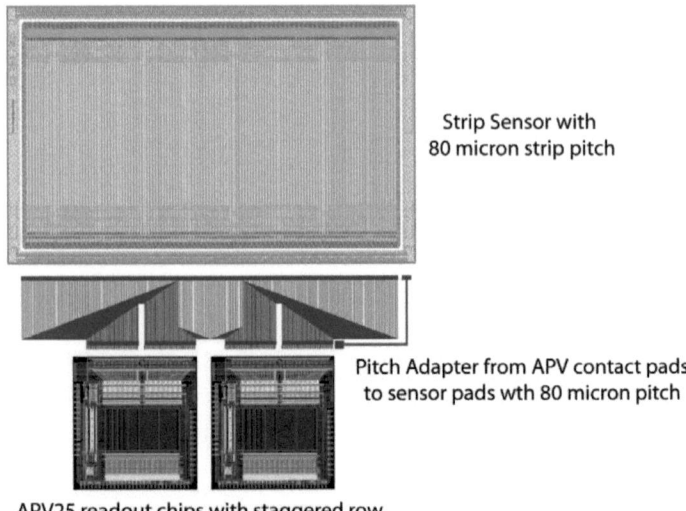

Strip Sensor with
80 micron strip pitch

Pitch Adapter from APV contact pads
to sensor pads wth 80 micron pitch

APV25 readout chips with staggered row
of contact pads with 44 micron pitch

Figure 8.1.: Illustration of the classic way of connecting readout chips to the sensors.

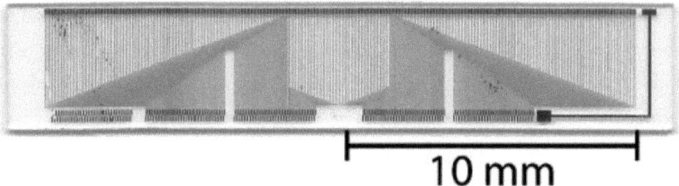

Figure 8.2.: A glass substrate with thin aluminium lines deposited on top of the substrate. It enables the connection of strips to the inputs of the readout electronics even if the pitch between strips and inputs is different, hence the name pitch adapter

8.1. On-Sensor Integration

The solution that we propose to solve the above mentioned problem, is the integration of the PA functionality into the sensor itself. The readout chips could be bonded directly to the sensor without the need of an external PA. This would have several advantages:

- The pitch adapter with its large and heavy glass substrate becomes redundant.
- Only one wire bond per channel is needed.

This results in a substantial reduction in material budget, less components during assembly, significantly reduced number of wire bonds and much more flexibility for the design of detector modules. Depending on the cost effectiveness of the on-sensor integration, a cost reduction can be achieved as well.

Three different approaches to integrate the PA on the sensor were prepared for this thesis:

Single Metal The metal layer which is used to form the readout electrodes of the strips can be used to route the strips to the contacts. The connection to the chip can only be located towards the edge of the sensor.

Double Metal An additional metal layer is deposited on the sensor, separated from the first metal layer by a thick oxide. The connection pads to the chip can be placed anywhere on the sensor.

Bump Bonding Uses the same double metal solution as the previous option but implements all necessary connections to bump bond an APV25 chip on top of the sensor, making the hybrid for the chips obsolete as well.

The first option is very cost effective, as it does not need any additional steps in the production while the second and third approaches require additional processing and therefore involve higher production costs.

8. Sensors with Integrated Pitch Adapters

In most cases, the metal lines used to route the signals to the connection pads will have to cross other adjacent strips, therefore introducing crosstalk between those strips. The first option will experience a larger coupling between such strips, as the oxide between metal and p+ strips has to be thin to achieve a high interstrip capacitance (C_{AC}, see section 2.2.2). For the second and third option, the thickness of the oxide between the metals is only defined by the requirements of the routing and can be chosen as thick as possible, usually limited by the capabilities of the manufacturing process.

As the *Single Metal* approach uses the metal layer of the readout electrodes to do the routing, there are limitations to the paths the lines can take. They cannot cross any of the readout electrodes, or the metallisation of the rings surrounding the strip area, or otherwise they will short-circuit them. The *Double Metal* option is much more flexible as the routing is in its own layer. Advanced sensors with a large number of strips and complicated requirements in routing might only be possible using the later option.

Due to the electrical advantages and the much higher flexibility, the *Double Metal* option is the preferred solution, while the *Single Metal* approach serves as a cost effective alternative for sensors with relaxed routing requirements.

The *Bump Bonding* approach is a very advanced concept which is presented here only as proof-of-concept. Bump bonding uses small metal balls to connect chip and sensor, where the full surface of the chip can be used to place the connection pads, in contrast to wire bonds which are limited to the periphery only.

One implementation of the process is described in [56]. It becomes clear that bump bonding usually requires special preparation of the bond pads on the sensor and the readout chip. The existing APV25 readout chips are only designed for wire bonding but certain bump bonding techniques could be modified to work on such pads as well [57].

8.2. Design of the Prototype Sensors

To evaluate the performance of single and double metal on-sensor PAs, a new production run at ITE Warsaw was envisaged. The wafer should contain five key structures (list includes the naming convention for each structure):

SensorST A standard AC coupled strip sensor as reference.

SensorPA Based on the layout of **SensorST** with an additional on-sensor pitch adapter in the first metal layer

Sensor2MPA Based on the layout of **SensorPA**, the pitch adapter is moved to the second metal layer while retaining its geometry as much as possible for easy comparison

8.2. Design of the Prototype Sensors

Sensor2MPABB Based on the layout of **SensorPA**, the pitch adapter is located on the second metal layer but designed to make bump bonding of an APV25 chip possible.

Halfmoon The full set of test structure should be included on the wafer as well.

8.2.1. Main Strip Parameters

The choice of the general strip parameters was more driven by boundary conditions as the influence on the performance of the integrated pitch adapters is rather small. We tried to optimise space allocation on the four inch wafers to accommodate the key structures while keeping some of the area available for an additional large sensors.

The design parameters for the strip sensors were narrowed down by the available readout electronics for beam tests. An obvious decision was the usage of the APV25 readout chip, as it is used in the original CMS Tracker and makes comparisons easier. Additionally, HEPHY has a lot of experience with the APV25 including a small and versatile readout system suitable for test beams[1] which was built by HEPHY's electronics department [58]. This fixed the number of strips on the sensor to a multiple of 128 AC coupled channels as provided by a single APV25 chip. As the main design parameters for the three key sensors should be kept identical and the usage of several readout chips per sensor would significantly blow up the space requirements on the wafer, the best compromise was to keep the number of strips at 128 or one single APV25 per sensor.

The strip pitch was decided to be 80 μm. This is the lower end of what is used in the innermost layers of the current CMS Tracker. The strip length was set to 40 mm to efficiently use the available space on the wafer.

8.2.2. Pitch Adapter Geometry

The main aim of the run was a proof-of-concept for on-sensor pitch adapters and the comparison between the two integration options. Considerable effort was spent to design a flexible geometric layout of the PA, including automated routing algorithms for SiDDaTA. Two concepts were devised, as seen in figure 8.3, which offer opposing advantages and disadvantages.

Shortest routing The routing lines were kept as short as possible using a direct connection between the strip and the contact pad. Not only the length of each routing line is depending on the strip number but the distance to its neighbouring routing lines as well. The crossings between the routing lines in the second metal layer and the strips in the

---

[1]The **APVDAQ** system is a prototype of the BELLE II SVD upgrade, also suitable for laboratory and beam tests.

8. Sensors with Integrated Pitch Adapters

first metal layer (or first metal and p+ implants for the single metal sensors) can have all possible angles. Control over crossing angles and routing line pitch is minimal and is mainly determined by the position along the strips were the routing starts

Constant routing line pitch The pitch between adjacent routing lines was kept at constant distance. Additionally, all crossings of routing lines and strips is rectangular. This precise control of the line pitch is counterbalanced by slightly longer routing lines. Even more problematic is that some routing lines run in parallel and very near to other strips. This could increase unwanted crosstalk among such strips.

8.2.3. Bump Bonding

To enable the bump bonding of an APV25, the chip has to be placed top down on the sensor. The second metal layer in the sensors was used to implement the very special routing scheme, which allows the bump bonding of the readout strips and the support lines of the chip as well. A technique developed at the Fraunhofer IZM [57] might be used to perform the direct bonding of the readout chip to the sensor, nevertheless in a first approximation the chip can also be glued to the sensor and connected using wire bonds.

Figure 8.4 shows a photo of the final sensor with an APV25 glued to the top of the sensor to illustrate the concept. The large pads on the sensor allow wire bonding for all readout channels and most of the support lines towards the right edge of the chip. The bump bonding method would require the chip to be placed top down on the sensor and the large pads would cover exactly the pad layout of the chip then.

Basic functionality tests have been done on the completed sensor confirming that the concept is working. Nevertheless, due to the same problems as will be reported in the subsequent section, the mechanical force of bonding procedure introduced many bad strips on the sensor making further investigations with this prototype difficult and out of the scope of this thesis.

8.2.4. Large Sensor with 512 Strips

The rest of the available space was used to implement a sensors with shorter but many more strips. The main idea was to design the sensor to be directly connected to an hybrid housing four AVP25 chips, as seen in figure 8.5, which was already available at the HEPHY in Vienna. The implemented routing layout should emphasise the influence of the integrated PA rather than being an optimal implementation. Therefore the routing was not centred on the sensor but moved to the far right.

8.2. Design of the Prototype Sensors

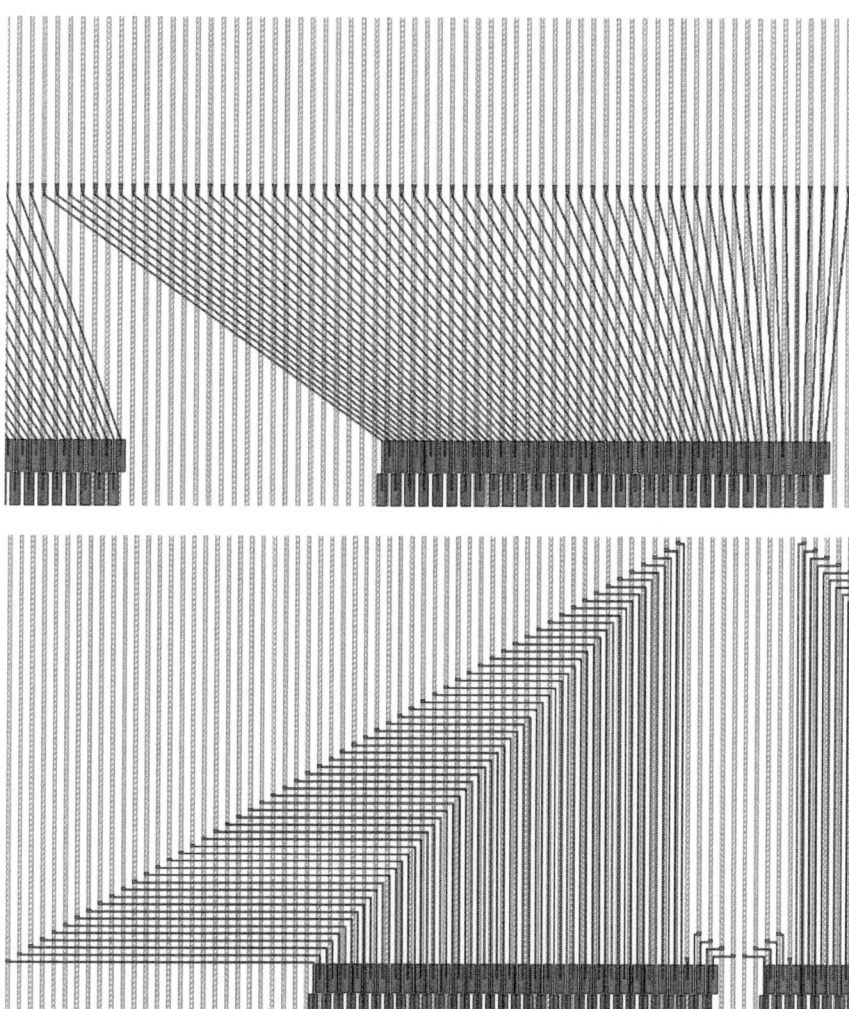

Figure 8.3.: Two concepts for the geometric layout of on-sensor PAs. The upper picture shows a part of the PA when trying to keep the routing lines as short as possible. Mark the varying inclination and the non constant pitch between neighbouring routing lines. The lower picture shows a part of the PA when trying to keep the routing line pitch constant and the crossings of strips and routing lines rectangular. Mark the long routing lines running in parallel to a different strip as seen towards the right of the picture.

8. Sensors with Integrated Pitch Adapters

Figure 8.4.: Photo of the final **Sensor2MPABB** with an APV25 readout chip glued to the top.

8.2. Design of the Prototype Sensors

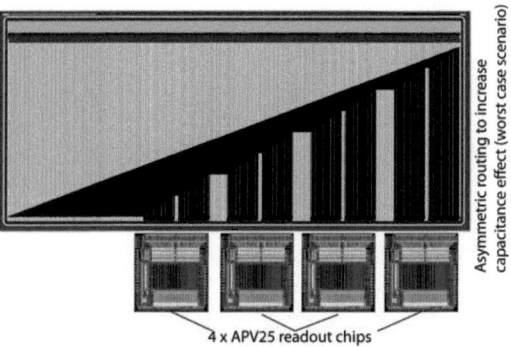

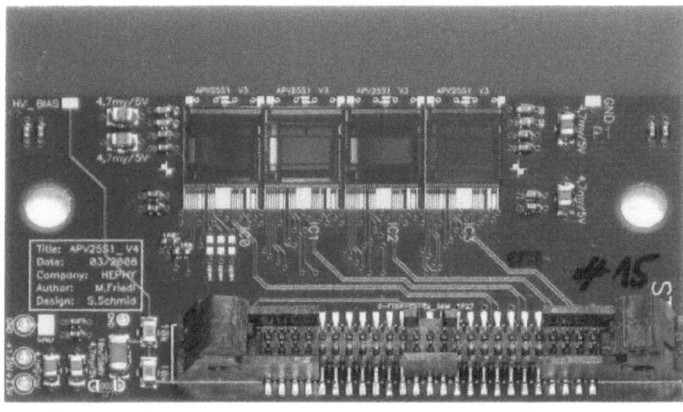

Figure 8.5.: Illustration of the concept of connecting four readout chips to one small sensor with very long routings (top) and a photo of the simple hybrid for four APV25 chips as it was available in-house at the HEPHY, Vienna (bottom).

8. Sensors with Integrated Pitch Adapters

8.3. Structure Summary and Wafer Layout

The target for the sensor designs were described in the previous section, while the actual implementations are shown in detail in this section. The drawings are screenshots of the actual mask layouts as produced by SiDDaTA and then used to manufacture the lithographic masks for the production.

Halfmoon with Test Structures including CapDM

The full set of test structures with all enhancements as explained in chapter 7 was included in the wafer layout. The new *CapDM* structure was included as well. Both are shown in figure 8.6

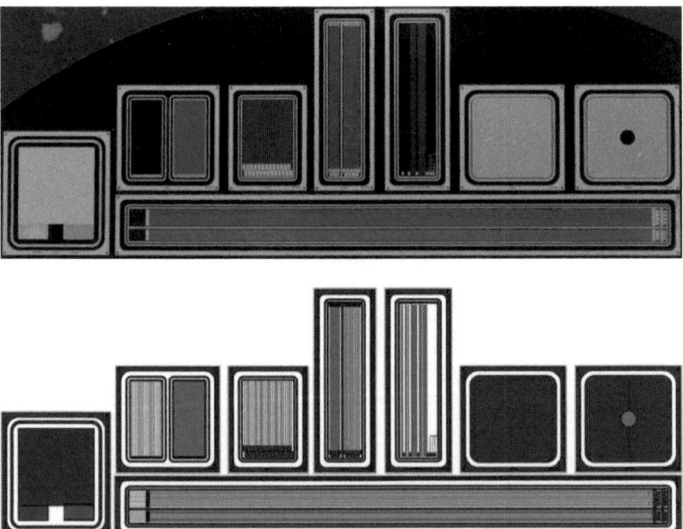

Figure 8.6.: Picture of the masks of the full set of test structure, extracted from the original mask design as prepared for the second run with ITE Warsaw (bottom) and a photo from the final structures cut from the wafer (top). From left to right: **CapDM**, **TS-Cap**, **Sheet**, **GCD**, **Cap-TS-DC**, **Diode**, **MOS** and at the bottom **Cap-TS-AC**.

8.3. Structure Summary and Wafer Layout

Standard Sensor: SensorST

The standard sensor design, as seen on figure 8.7, on which the following three sensors are based on. It has 128 AC coupled strips with polysilicon resistor biasing specified at 10 MΩ. The strips are 40.1 mm long and 20 μm wide with a pitch of 80 μm. The metal overhang is 5 μm at each side and the strip area is surrounded by a bias and a guard ring with an asymmetric metal overhang. A 500 μm wide n+ implant protects the edge of the sensor.

Figure 8.7.: The standard sensor design **SensorST**.

Single Metal PA: SensorPA

Using the same main strip parameters as **SensorST**, this sensor uses the first metal layer to implement the on-sensor PA. The metal strips, normally used as the upper readout electrode which couples capacitively to the strip implant, is converging at one side to the pitch of the APV25 chip. Figure 8.8 shows the PA section of the sensor design.

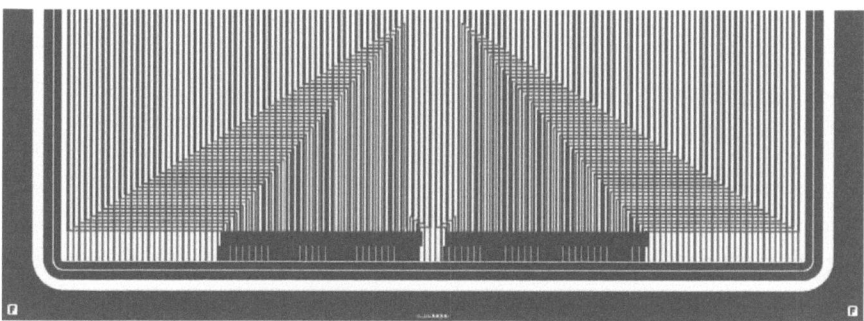

Figure 8.8.: Section of the single metal pitch adapter design **SensorPA**.

153

8. Sensors with Integrated Pitch Adapters

Double Metal PA: Sensor2MPA

Based on the same design parameters as **SensorPA**, the PA is now located in the second metal layer as seen in figure 8.9. The geometry of the PA is kept exactly the same as in **SensorPA** where vias at the end of each line connect the routing lines to the readout strips. The second metal routing lines are kept as short as possible.

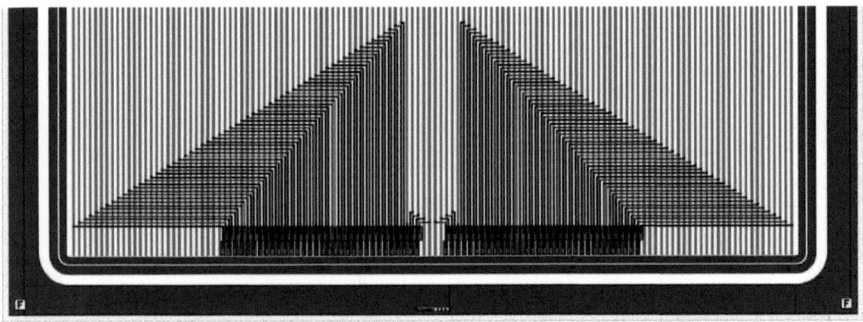

Figure 8.9.: Section of the double metal pitch adapter design **Sensor2MPA**.

Double Metal PA for Bump Bonding: Sensor2MPABB

This sensor is again based on the same main strip parameters as in **SensorST** but implements a very sophisticated connection scheme in the second metal layer as seen in figure 8.10. It allows the possibility to bump bond an APV25 chip up-side-down to the sensor. The PA layout does not only include the necessary connections to the readout strips, but also the necessary backend connections of the readout chip, including a 50 Ω resistor which taps the 2.5 V line and supplies 1.5 V. As a first, more conventional approach, the APV25 will be glued to the sensors and connected with wire bonds as already shown in the previous section 8.2.3.

8.3. Structure Summary and Wafer Layout

Figure 8.10.: The special double metal pitch adapter design **Sensor2MPABB** for APV25 bump bonding.

Double Metal PA for 4 APV Hybrid: Sensor2MPA90

This sensor differs in some of the main strip parameters as it has $4 \times 128 = 512$ strips which are shorter than in the other sensor at 18.1 mm. The routing scheme in the second metal layer is

8. Sensors with Integrated Pitch Adapters

therefore a little more complex with the additional feature that it is not centred as seen in figure 8.10. This is not the ideal case as it increases the length of the routing lines for almost all strips. Nevertheless, this was chosen to increase the influence of the routing and reflect scenario which is more akin to a real sensors with even more strips and higher integrated readout chips.

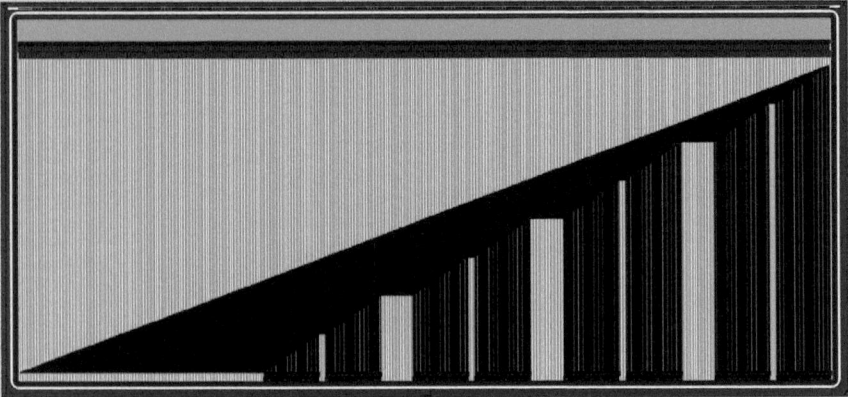

Figure 8.11.: The **Sensor2MPA90** has somewhat different strip parameters than the previous sensors in terms of number and length but is otherwise very similar. The pitch adapter is not centred.

Full Wafer Layout

Figure 8.12 shows the layout of the completed wafer. The **Halfmoon** is placed in three locations towards the edge of the wafer, while the **CapDM** structure was placed at four locations. The five sensors occupy most of the space towards the centre.

8.3. Structure Summary and Wafer Layout

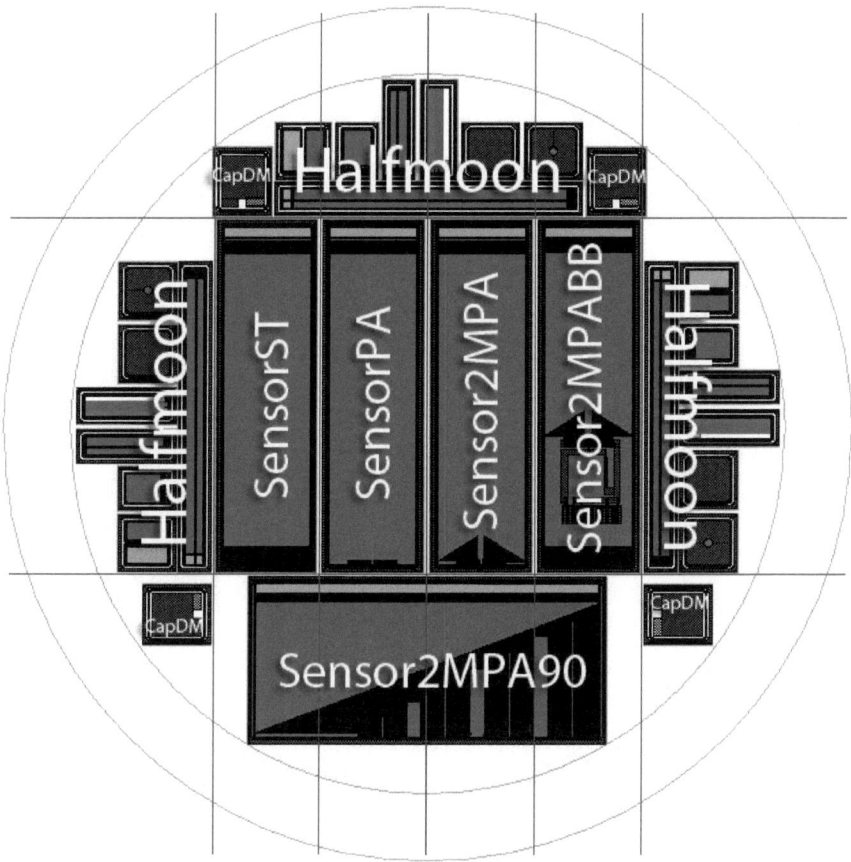

Figure 8.12.: The full layout of the wafer.

Process Parameter

The process parameters were determined as a compromise between the optimal choice for the performance of the sensor and the capabilities of the potential producers for large volume production available on the market. Table 8.1 quotes the most important process parameters.

The process was implemented in two slightly different sequences to produce single and double metal wafers with only slight changes. Ten high resistive four inch wafers were used

8. Sensors with Integrated Pitch Adapters

for the production. Up to the first metallisation the process sequence was exactly the same for both single and double metal. Wafers 1 to 5 then received their final passivation resulting in wafers with one single layer of metal. For the other wafers 6 to 10 an additional silicon oxide insulation was added, vias were etched into it and the second metal layers was deposited on top. The final passivation used the same mask layout as for the single metal design. For all single metal designs additional care was taken, to allow access to all contact pads in the first layer.

Thick interstrip oxide:	900 nm
Thin readout oxide:	177 nm
PolySi:	473 nm, target resistivity 6.8 kΩ/sq
Insulation PolySi - 1st metal:	thermal 43nm, CVD SiO2 100nm, BPSG 700nm, PSG 200nm
Insulation 1st - 2nd metal:	PE CVD SiO2 700nm
1st metal:	Alu-Si 700 nm
2nd metal:	Alu-Si 1200 nm
Passivation:	PSG 450 nm, AP CVD SiO2 300 nm
Backside Al:	20 nm (very thin, to allow laser tests)
Backside getter:	1000 nm PolySi
Backside n++ implantation:	6-8 μm

Table 8.1.: The most important parameters of the production process. Indicated values are the ones used to setup the process and not the measured results of the finished products.

From the ten wafers which were processes, we received eight wafers, where wafers 1, 2, 3 and 4 were only processed with a single metal layer and wafers 5, 6, 7 and 8 received the full double metal processing. Wafer 10 broke during manufacturing and wafer 9 was not sent due to very high reverse bias currents. According to the ITE Warsaw, this was due to not applying a getter procedure at the backside of the wafer. Additionally, for wafer 4 the **Sensor2MPA90** was damaged and was not sent to vienna including the two CapDM structures on the same cutoff.

8.4. Electrical Characterisation

The test structures and sensors were electrically characterised in the clean room at the HEPHY in Vienna. Two probe stations were used to make the measurements, where one station was suitable to perform a full semi-automatic strip characterisation (*QTC Station*, see [59]), while the test structures where measured in slightly different setup without a motorized table (*PQC Station*, see [28]).

As explained in chapter 7, the measurements on the test structures enable the judgement of the quality of the manufacturing process. Parameters like the full depletion voltage, oxide thicknesses or material resistivities are checked to comply to the specifications. In section 7.3, the measurements on the test structures are already documented and the results are used to asses the correct design of the structures themselves. In this section we review the results again to judge the quality of the manufacturing process.

The sensors themselves were also characterised by measuring the most important strip parameters and global IV curves. Sensors were generally tested in the process version which fits the sensor layout. For the single metal sensors (**SensorST** and **SensorPA**) wafers 1 to 4 had the appropriate single metal processing, while for the double metal sensors (**Sensor2MPA**, **Sensor2MPA90**) wafer 5 to 8 had the fully compatible processing. To check if the additional process steps of the double metal process have a general impact on sensor performance, we tested selected single metal sensors and test structure on wafers 5 to 8 as well. A special effort was made to identify the electrical modifications introduced by the PAs and to disentangle them from effects solely caused by fluctuations in the manufacturing process.

8.4.1. Process Quality Control

The results from the measurements on the test structure are presented in section 7.3 and only the relevant results to judge the process quality will be discussed here.

The measurements from the Diode structures show a reasonably low reverse bias voltage and high voltage stability up to and beyond 1000 V. They all show some instabilities above 1000 V while some diodes even experience early breakdown. Nevertheless, the quality should be sufficient to fully deplete the sensor devices as the full depletion voltage, extracted from the CV-curves of diodes, is far below at around 24 V.

The thickness of the oxide between the two metal layers can be measured with the CapDM structure. The result of about 600 nm μm as presented in table 7.4 agrees well with the specified parameter of 700 nm of SiO_2 (before reflow) as seen in table 8.1.

The resistivities of the polysilicon in the Sheet structure as seen in table 7.6 deviates significantly from the target resistivity of 6.8 kΩ/sq as shown in table 8.1. This should influence the resistance of the polysilicon resistors in the sensors and should be seen in the stripscan performed in the subsequent section.

A more serious issue is the thickness of the thin readout oxide between p+ strip implant and

8. Sensors with Integrated Pitch Adapters

the metal readout strip. It should be only a thin layer to maximise the capacitive coupling. According to table 8.1 it was specified at < 200 nm while the measurements on the CapTS structure yielded a thickness of around 600 nm.

8.4.2. Sensor Quality Control

An electrical characterisation was done for all of the sensors prior to integration and bonding of the modules. The main goal was to select the sensors which could be safely operated above full depletion and which had only a small number of bad strips.

Reverse Bias Current - IV

The sensors showed a varying quality in terms of reverse bias current even on a single wafer. Some general conclusions can be drawn despite the low statistics offered by the eight finalised wafers as seen in figures 8.13 and 8.14. Most of the sensor can be safely operated beyond full depletion and wafers with double metal processing seem to be equally stable than wafers with single metal processing. Also the type of sensor does not deteriorate the voltage stability of the sensor, except for the larger **Sensor2MPA90**. This is to be expected, as the sensor does not only occupy much larger surface, but during processing it was located towards the edge of the wafer. This increases the chances that the sensor includes small flaws of the manufacturing process and therefore limits its stability.

The sensor design itself is high voltage robust up to and far beyond 200 V as figure 8.15 shows. The sensor exhibits an almost ohmic behaviour beyond 200 V which is typically caused by instabilities in the processing and not by a general flaw in the design. Assuming a stable production process, the sensor design alone should be stable at even higher voltages without causing breakthrough.

Full Depletion Voltage - V_{FD}

Capacity – Voltage curves were performed on several sensors of different wafers and sensor types. The interesting parameter derived from such curves is the full depletion voltage V_{FD} as explained in section 1.5.1. The onset of the plateau defines V_{FD} and is consistent for all structures shown in figure 8.16 at about 25 V. This is as expected for a parameter which depends only on the wafer material where 25 V corresponds to a bulk resistivity according to equation

8.4. Electrical Characterisation

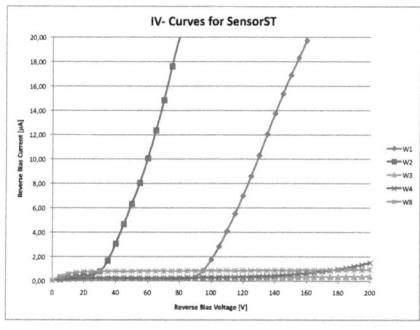
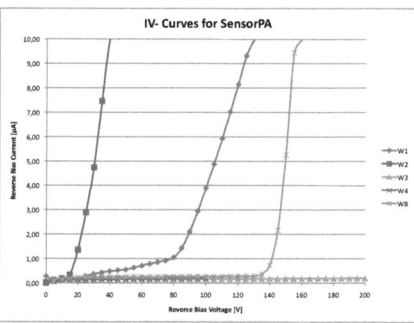

Figure 8.13.: IV-curves for the single metal sensors on wafers with single metal processing (W1 to W4). Additionally both sensor types were also tested on wafer 8 which received the full double metal processing.

1.65 of:

$$\rho = \frac{d^2 + 0.6322 pd}{2\varepsilon_0 \varepsilon \mu_n V_{FD,sensor}} = \frac{(300 \times 10^{-6})^2 + 0.6322 \cdot 80 \times 10^{-6} \cdot 300 \times 10^{-6}}{2 \cdot 8.85 \times 10^{-12} \cdot 12 \cdot 0.135 \cdot 25} \quad (8.1)$$

$$\rho = 147 \ \Omega m = 14.7 \ k\Omega cm \quad (8.2)$$

This fits the specifications of the wafer supplier which were (vaguely) defined as $\rho > 10 \ k\Omega cm$.

Stripscans

For all sensors a stripscan was performed were four parameters were measured per strip: the strip current I_{strip}, the PolySi resistance R_{poly}, the interstrip resistance R_{int} and the current through the dielectric I_{diel}. Several problems were identified and they are exemplary shown for the **SensorST** layout.

The current per strip shows no apparent anomalies as seen in figure 8.17. The average current per strip is around 1-2 nA and for some sensors a characteristical gradient from leftmost to rightmost strip is seen. Some strips do experience a higher reverse bias current but this was to be expected for such a first prototype run.

The ohmic resistance of the polysilicon resistors were specified at 10 MΩ. The results showed a significantly lower value on all sensors and wafers at around 0.2 to 0.4 MΩ as exemplary shown in figure 8.18, with slightly higher values on wafer 4 only.

A common problem for AC-coupled sensors are strip implants which are shorted to the readout aluminium due to bad or broken oxide between them. Such shorts, usually called pinhole,

8. Sensors with Integrated Pitch Adapters

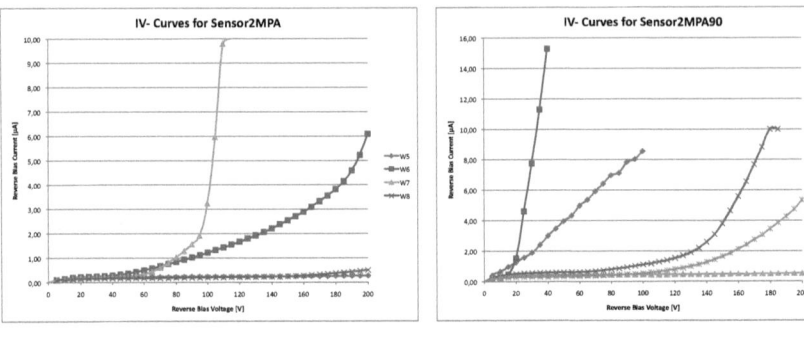

Figure 8.14.: IV-curves for the double metal sensors on wafers with double metal processing (W5 to W8). **Sensor2MPA90** on wafer 7 showed break through already at 1 V. Additionally **Sensor2MPA90** was also tested on wafers 2 and 3 which only received the single metal processing.

would cause the reverse bias current of the strip to flow through the preamp of the readout chip. As the main purpose of the built-in strip capacitor is to prevent this, the readout chip, such as the APV25, cannot cope with this increase in current and could be rendered inoperable by a few of such defects. The stripscan is able to identify such pinholes by measuring the current through the oxide: a small voltage is applied between DC pad (connected to the strip implant) and the AC pad (connected to the readout strip above) and the current is measured. The coupling or readout capacitance is also measured between implant and readout strip.

In figure 8.19 both results are exemplary plotted for several sensors of different wafers. In the order of 10 - 20 pinholes (corresponding to 10 to 20% of the strips per sensor) are identified for each sensor among all wafers and sensor layouts. This again hints to problems in the manufacturing of the thin readout oxide. While in the previous section the coupling capacitance measurements on the diode showed a significantly thicker oxide then expected, the pinhole measurements suggest in addition a poor mechanical quality of the oxide.

Summary and Discussion with ITE Warsaw

Several problems were discovered by the electrical characterisation of the sensors and the test structure:

- The polysilicon showed a significantly lower resistivity, confirmed by the low resistance of the bias resistors on the sensors.
- The thickness of the readout oxide between strip implant and readout strip is much thicker than specified.

8.4. Electrical Characterisation

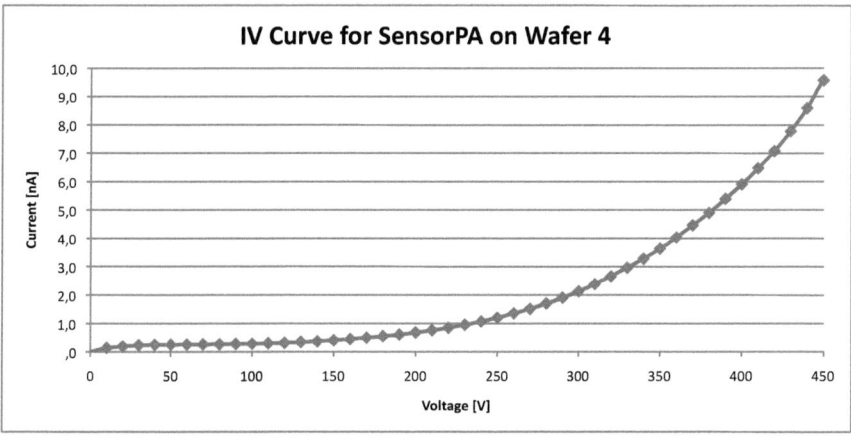

Figure 8.15.: IV-curve for **SensorPA** on Wafer for up to 450 V.

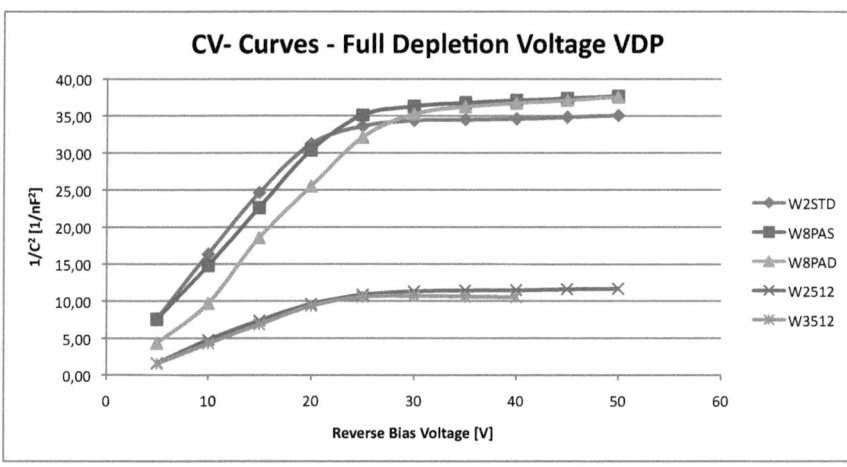

Figure 8.16.: CV curves for several structures on different wafers produced in the 2009 run with ITE Warsaw. The capacity is plotted as inverse square to emphasize the knee in the curve. The final sensor capacity is similar for the sensors with the same overall layout (**SensorST**, **SensorPA** and **Sensor2MPA**) while the slightly larger **Sensor2MPA90** show a consistently higher capacity.

163

8. Sensors with Integrated Pitch Adapters

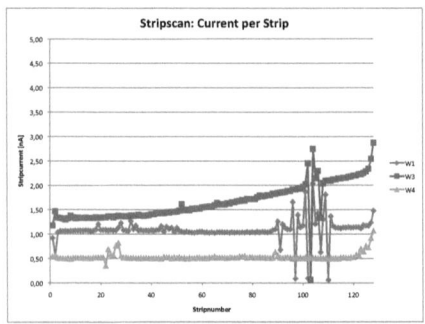 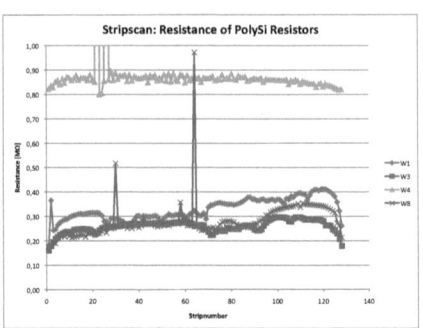

Figure 8.17.: Reverse bias current for each strip at 50 V. Only the **SensorST** structure is exemplary shown, as the behaviour for all sensor designs and wafers is similar and shows no apparent problems.

Figure 8.18.: Resistance of the polysilicon resistor of each strip which was specified at 10 MΩ. While all sensors on all wafers showed a similar behaviour, **SensorST** on wafer 4 exhibited slightly higher values.

- The mechanical quality of the readout oxide is poor as suggested by the high number of pinholes ($\approx$ 10-20%).

In the discussions with the supplier ITE Warsaw, all the reported observations could be traced back to the responsible source in the manufacturing process. The low resistivity of the polysilicon was pinpointed to wrong settings during the doping of the material. This process is very sensitive to small changes in the process parameters and have to be tuned to the appropriate and stable settings from the experience with the actual production.

The wrong thickness and poor quality of the thin oxide between p+ implant and metal is caused by the production process implemented by ITE Warsaw. This process sequence, referred to as *polysilicon gate CMOS*, was derived from a standard process used for CMOS production where a sandwich of (thin) oxide, polysilicon and (thick) oxide is located between metal and p+ implant. By omitting the polysilicon between p+ strip and readout metal, the oxide is composed of two layers which greatly increase the thickness of the oxide while degrading its quality. For a future run, scheduled to be delivered in summer 2010, a modified process sequence will be used which should enable the creation of only a single thin oxide layer between p+ implant and the metallisation. This process is referred to as *aluminium gate CMOS*.

Such iterations in the optimisation of the manufacturing process are to be expected. Nevertheless, the electric characterisation of sensors and test structure again proved to be an indispensable tool to identify and trace problems and modify and improve the manufacturing process accordingly.

8.4. Electrical Characterisation

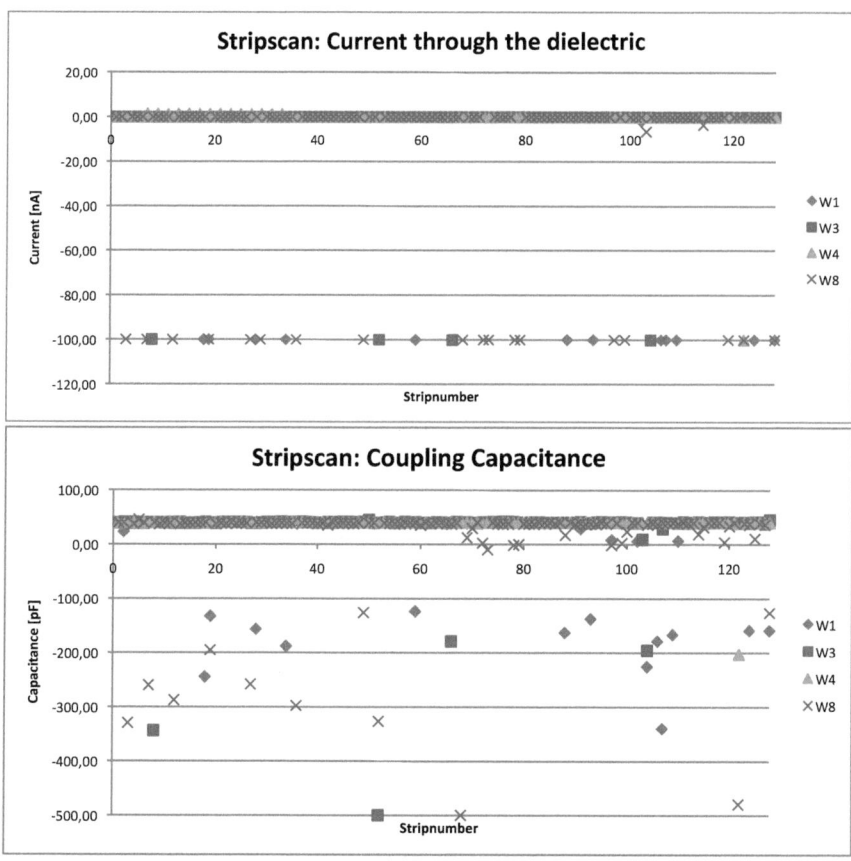

Figure 8.19.: Stripscans of the current through the dielectric (upper) and the coupling capacitance between readout strip implant (lower). A high number of pinholes is identified in the upper plot for strips with a high (negative) current. The maximum current is limited by the power supply to 100 nA. The lower plot confirms the pinholes for strips identified in the upper plot. The negative values for the coupling capacitance are an artifact of the measuring equipment, trying to define a capacity for shorted contacts. All values smaller than zero are effectively zero.

8.5. Beam Tests at CERN's SPS

The sensor were field-tested in a low intensity 120 GeV/c Pion beam at CERN's Super Proton Synchrotron (SPS). We used the MIP like particles in the beam to asses the performance penalty introduced by the on-sensor PAs.

8.5.1. Module Construction

The sensors were integrated in robust modules housing a small hybrid with one to four APV25 readout chips and a connector to the readout system as seen in figure 8.20. The mechanical support structure was milled from a fibre-reinforced epoxy material called *Isoval 11*, which provides good mechanical stability while it is easy to machine. Openings in the frame let the beam pass unobstructed but they had to be covered with black adhesive tape to prevent the exposure of the sensors to light during operation.

The APV25 readout chips were always bonded directly to the sensor. This is the natural choice for the sensors with integrated PAs, but for a better comparison of the sensors themselves, this was also done for the standard sensors **SensorST**. By exercising some caution during bonding, it is possible to connect a single APV25 chip to the larger 80 μm pitch of the strips.

Figure 8.20.: Modules built for the 128 channel sensors (left) and the 512 channel sensors (right). The small print to the right of the modules houses the APV25 readout chip(s) and a connector.

The modules were constructed in such a way, that several of them could be easily combined to larger stacks as seen in figure 8.21. This increased the data collected per event, while multiple scattering was negligible for 120 GeV/c particles.

The wire bonding of the sensors with the PA on the second metal layer proved to be problematic. On all of these sensors the number of pinholes measured before and after bonding

8.5. Beam Tests at CERN's SPS

Figure 8.21.: A stack of several sensor modules. The stack contained not only the sensors described in this chapter but also some additional references sensors. Some of these sensors were rotated by 90° which means that the strips were oriented perpendicular to the strips of the ITE sensors. This provided height information for each hit.

increased significantly, rendering most of them unusable. It is well known, that pinholes can be induced by the mechanical force applied to the connection pad during bonding. This usually only happens due to improper settings of the bonding machine.

Figure 8.22 shows the connection area of a **Sensor2MPA**. It is clear, that the composition of materials below each pad is different depending on the strip number. The pad can be situated between two strips and therefore over the thick interstrip oxide, or it might be directly over a strip on the thin readout oxide. The transition between the thin readout oxide above the strip to the thicker oxide in the interstrip region creates a step in the overlying pads which was suggested by the manufacturer as being a weak spot to mechanical stress. Nevertheless, traditional connection pads on the first metal layer are also located above the thin oxide and extend into the thick oxide of the interstrip region. As such pads can be bonded without problems, the reduced mechanical robustness is more likely an artifact of the processing of the second metal layer and the insulating layer between the metal layers.

As already mentioned in the previous section, the modified *polysilicon gate CMOS* process used by ITE Warsaw to manufacture the sensors, was not the standard choice for strip sensors. It produced a low quality oxide between p+ implants and metal strips, which might be further weakened by the additional processing steps necessary for the second metal layer. Improvements to the newly suggested and in general more suitable *aluminium gate CMOS* process are currently discussed, involving reflow steps to smooth sharp steps between thick and thin oxide and thicker low temperature oxides between the two metal layers.

8. Sensors with Integrated Pitch Adapters

Figure 8.22.: Close-up of the contact pad region of a **Sensor2MPA** sensor. The location of each pad with respect to the strips depends on the strip number. Picture was made using a reflected-light microscope.

8.5.2. Readout Electronics and Services

The off-detector system used to readout the APV25 chips is a prototype developed at the HEPHY for the upgrade of the BELLE II Silicon Vertex Detector at the KEKB Accelerator in Tsukuba, Japan. The system offers advance features like precise hit time finding by taking six samples for each trigger using the APV25's multi peak mode [60]. More importantly, the prototype system was capable of controlling and reading the large number of APV25 chips that were necessary to read out the stacks composed of several modules.

The high voltage needed to deplete the sensors was supplied by standard laboratory power supplies which included the monitoring of the current. We choose a common reverse bias voltage of 70 V for all sensors, which is far above the full depletion voltage of the sensors (around 25 V). Temperature and relative humidity was measured during all runs to ensure them to be within uncritical values (T = 25± 3°C, RH < 50 %)

8.5.3. The Beam

We utilised the H6A beam area at CERN's North Area Hall. The protons supplied by the SPS accelerator complex were converted using a target and tuned to low intensity at 120 GeV/c momentum. The final beam arriving at the test area had a composition of 55.67 % Π^+, 38.95 % protons and 5.38 % K^+. Following the time structure of the protons supplied by the SPS, the H6 beam line receives a few seconds of protons during a full SPS cycle of roughly 20 seconds. Each spill provided around 3×10^5 particles. In figure 8.23 the beam profile of a single spill is displayed, where the size of the beam spot was roughly constant among all the runs taken (horizontal: around 16 mm, vertical around 7mm).

8.5.4. The Test Setup

The test setup is shown in figure 8.24. We utilised the EUDET Beam Telescope [61] which already provided a stable support platform including a motorised table with X-Y-Z and rotational controls. The trigger was produced by an array of two scintillators before and two past the telescope detector planes. Trigger information was sent to the readout of the EUDET Beam Telescope and the readout system of our Device Under Test (**DUT**) including a timestamp. While the telescope was not necessary for the analysis done for this work, the data was recorded and is available on disk for later reference.

The test beam was carried out between 19. - 26. August 2009 were we collected more than 3 million events. The data was not only collected with the sensors described in this work, but a number of other sensors were tested as well.

8. Sensors with Integrated Pitch Adapters

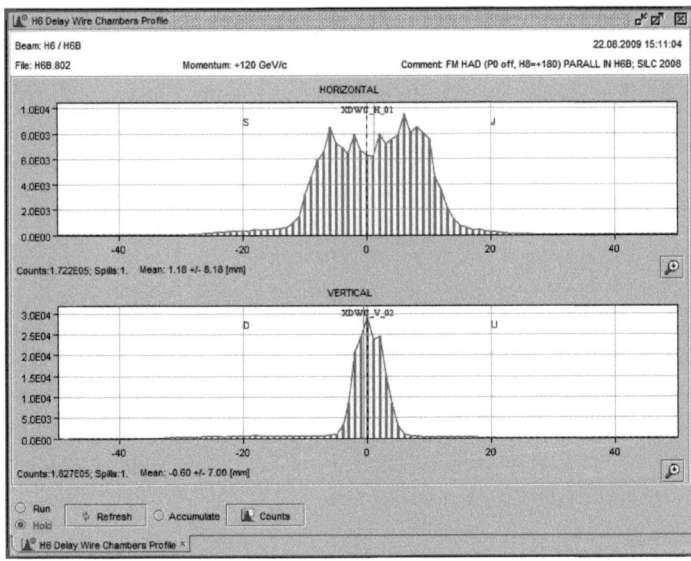

Figure 8.23.: The beam profile as measured with wire chambers located close to the device under test. The measurement was taken from a single spill.

Figure 8.24.: Test setup as used during the testbeam. The two boxes with the **DESY** logo each contain three planes of pixel sensor for the telescope. The beam enters the setup from the left.

8.5.5. Results

Regarding the performance of sensors with integrated PAs, the two most important issues are the possible deterioration of the signal-to-noise ratio and the amount of additional cross talk introduced by the routing. Both issues are influenced by several parameters and have to be dealt with individually for single and double metal designs.

For all upcoming plots, the calibration of signal to electrons is done using the APV25 built-in calibration pulse method [62], [63] and [64]. The chip has the capability to create an internal calibration signal of known charge and inject it on the strip. The signal measured on the strip can than be calibrated to the known number of charge carriers. The method has a limited precision and gives only a coarse calibration.

Signal and Noise

The signal induced by an incident particle has to travel along the additional routing line. As these lines are rather narrow (8 µm) and therefore have a larger resistance than the wider (30 µm) strip metallisation, the signal will deteriorate. Moreover, the additional capacity introduced by the routing line could increase the capacitive noise picked up by the readout chip. Usually, both effects are also found on traditional chip to sensor connection using external PAs,

8. Sensors with Integrated Pitch Adapters

where the added contributions are caused by the PA itself and the second bond necessary to connect to the sensor.

As described earlier in section 8.5.1, all modules have an immediate connection between readout chip and sensors using wire bonds. This is also true for the standard sensor **SensorST** to offer a better direct comparison of the sensors performance alone.

The sensors with a single metal PA should experience an additional effect due to the missing strip metallisation in the region of the integrated PA as seen in figure 8.25. As the metal is used to route the signal to the readout pads, part of the strip is missing its metallisation. In a naive way of thinking, signals created by particles inside the PA area have to travel along the uncovered strip before they are capacitively linked to the metal strip. As the resistance of the p+ implant is much higher than in the aluminium, some loss of signal can be expected.

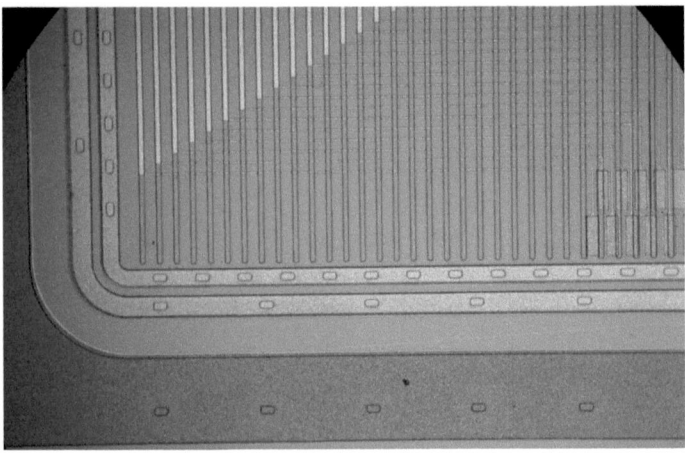

Figure 8.25.: Close-up of the pitch adapter region on the single metal sensor **SensorPA**. The metalisation over the strips is used to route the signal to the readout pads.

Figure 8.26 shows a comparison of the noise on a sensor with different PA concepts. For the single metal sensors, the noise distribution is inconspicuous with a slight tendency of more strips with high noise for the **SensorPA** structure. Nevertheless, there is no evidence for an increase in noise caused by the additional capacitance of the routing strips.

The double metal sensor shows many strips with extremely high or low noise, originating from bad channels. As already described in section 8.4, the low quality oxide encouraged the creation of bad channels during bonding of the sensor. The distribution of bad channels is flat

8.5. Beam Tests at CERN's SPS

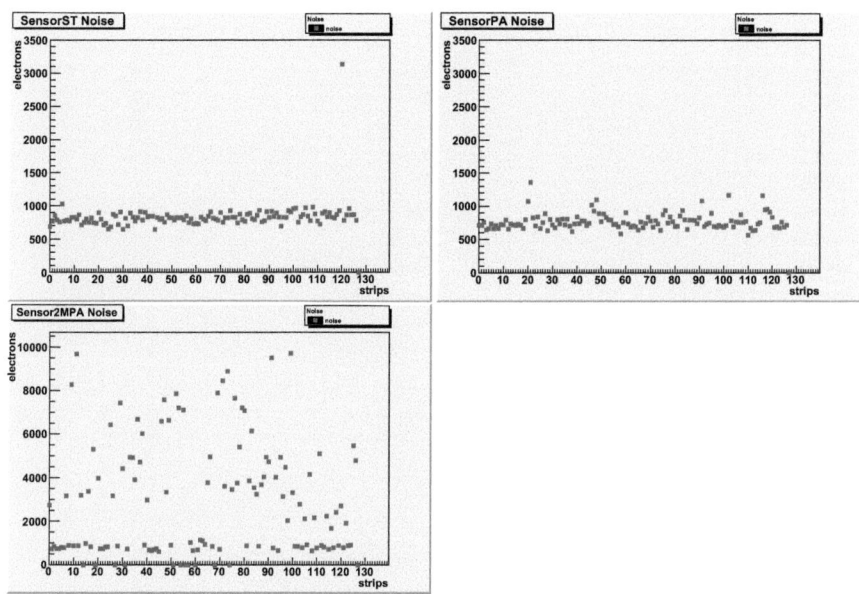

Figure 8.26.: Noise on a standard sensor without pitch adapter (**SensorST**, top - left), on a sensor with single metal pitch adapter (**SensorPA**, top - right) and on a sensor with pitch adapter in a second metal layer (**Sensor2MPA**, bottom - left). The noise is calibrated to electrons using the APV25 built-in calibration pulse method. The non-uniform noise distribution on the **Sensor2MPA** is caused by bad channels.

over the full width of the sensor, although the bonding pads are only located over strips 28 to 102. This indicates, that the mechanical force applied during bonding did damage the oxide between the two metal layers and between p+ implant and first metal. The former would cause an ohmic connection between different strips resulting in a higher capacity and therefore increased noise, while the later would cause a short-circuit of the readout capacitor and therefore reduce the capacity and the noise.

Unfortunately, this makes the interpretation of the results from double metal sensors very difficult. Various unknown artifacts in the data can be created by the bad strips, distorting the upcoming plots on **Sensor2MPA**.

The signal created by particles hitting the sensor is shown in figure 8.27. It is defined as the most probable value of a convoluted Landau-Gauss fit on the signals from all hits on a certain strip (see section 2.1). The expected signal created by a MIP-like particle in 300 μm of silicon is 23.000 electron-hole pairs as discussed in section 2.1. The standard sensor **SensorST** fits

8. Sensors with Integrated Pitch Adapters

these expectations quite well, taking the coarse calibration method into account.

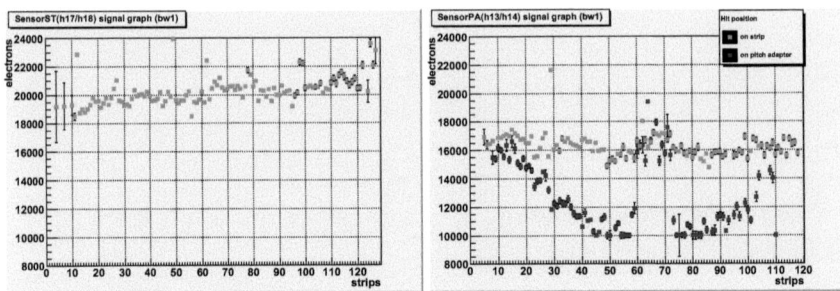

Figure 8.27.: Signal on a standard sensor without pitch adapter (**SensorST**) and on a sensor with single metal pitch adapter (**SensorPA**). The plot for **SensorPA** has signals calculated for hits inside the area of the pitch adapter (dark) and outside (light). The signals are calibrated to electrons using the APV25 built-in calibration pulse method.

For **SensorPA**, the signal created by hits in the area outside the PA tend to be lower than for **SensorST**, which is probably an artifact of the calibration method. More important is the loss of signal for hits inside the PA area. Depending on the length of the uncovered area of the strip, the loss in signal can be up to 50%.

Naturally, the same behaviour can be found in the signal-to-noise plots in figure 8.28, where each signal is normalised to the noise of the hit strips. Here, the signal-to-noise ratios for **SensorST** and **SensorPA** hits outside the PA are equal to 20. This confirms the suspicion, that the lower signals for **SensorPA** in figure 8.27 are just an artifact of the calibration method.

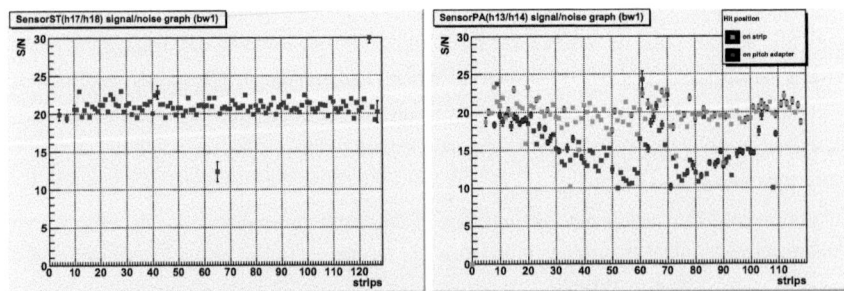

Figure 8.28.: Signal-to-noise ratio (S/N) on a standard sensor without pitch adapter (**SensorST**) and on a sensor with single metal pitch adapter (**SensorPA**). The plot for **SensorPA** has signals calculated for hits inside the area of the pitch adapter (dark) and outside (light). The signals are calibrated to electrons using the APV25 built-in calibration pulse method.

Crosstalk

Due to the capacitive coupling between the routing lines and the strips p+ implant (in the single metal designs) or the strips metallisation (in the double metal designs) some of the signal induced in the strip is shared with other strips. The crossings of the routing lines and other strips are one source of crosstalk, but as the area of these overlaps is rather small, the effect should be small as well. A naive calculation using a simple parallel plate capacitor model and approximating single and double metal designs with the same cross-section yields:

$$C_{crossing} = \varepsilon_0 \varepsilon_{ox} \frac{w_s w_r}{d} = 8.2836 \times 10^{-15} F \approx 8 \text{ fF} \quad (8.3)$$

where the dielectric constant of vacuum $\varepsilon_0 = 8.85 \times 10^{-12} F/m$, the dielectric constant of silicon dioxide $\varepsilon_{ox} = 3.9$, the width of the metal strips $w_s = 30$ μm and the width of the routing lines $w_r = 8$ μm and the thickness of the dielectric is $d = 1$ μm.

A more pronounced effect can be expected, if routing lines run in parallel and near to or fully overlapping a different strip. The same naive calculation as above, assuming the worst case were the routing line fully overlaps the strip, yields:

$$C_{ll,overlap} = \varepsilon_0 \varepsilon_{ox} \frac{l_u w_r}{d} = 2.7612 \times 10^{-13} F \approx 280 \text{ fF/mm} \quad (8.4)$$

where l_u is a unit length of one mm. The longest routing line directly overlapping a strip is 3 mm resulting in a capacity of already 740 fF.

To get an overview of the crosstalk introduced by the routing in the first and second metal layer, we calculated correlations using the Kendall tau rank correlation coefficient [65]. For an event where a certain strip was hit, we compare the signal height on all other strips (the strips not hit in this event) to the signal heights of the same strips in all other events where the identical strip was hit. If the difference in signal height between two events has the same sign as the difference in signal height of the strip that was hit, we count it as a concordant pair (n_c) and as non discordant pair (n_c) if the sign is negative. For each pair of hit strip (X-Axis) and non hit strip (Y-Axis) we get Kendall's tau which is defined as:

$$\tau_K = \frac{(n_c - n_d)}{0.5 n(n-1)} \quad (8.5)$$

The rank correlation coefficient τ_K is color coded and plotted for each pair of strips in figure 8.29 for a sensor with a PA on the first metal layer (**SensorPA** and one without (**SensorST** as comparison. τ_K is very sensitive to even very weak correlations, therefore pointing out any

8. Sensors with Integrated Pitch Adapters

causes for even a minimal crosstalk. Ignoring the various horizontal and vertical patterns which are caused by bad strips, two diagonal lines are observable which are symmetrical around the 45° line of identical strip pairs. This line directly corresponds to the geometrical pattern of routing on the sensor as seen in figure 8.8.

While the Kendall tau rank coefficient reliably points out even very weak correlations, it does not give any measure of the strength of the coupling. A second method was used to determine how much of a signal originating from a strip hit by a particle can be seen on a different strip which was not hit. For a strip which was hit by a particle, corresponding to a signal which was above the threshold of 5 σ of the strip noise, its signal is plotted against the signal measured on a second strip at the same time, resulting in a plot like the example in figure 8.30. The slope is determined by calculating the mean of all hit-to-other strip relations. This *gainfactor* is then color coded and plotted for each pair of strips in figure 8.32 and 8.33.

What is clearly seen in the lower gainfactor plot for **SensorPA** in figure 8.33, are the same lines caused by the geometry of the on-sensor PA as seen in figure 8.29. The line with the steeper slope is more pronounced than the line with the shallow slope. The steep slope is caused by signals which are more likely deposited on a strip and directly coupled to a routing line running in parallel to the strip implant as sketched in the left drawing in figure 8.31, where for the shallow slope the opposite is more likely, a signal is deposited on a strip then runs along its routing line and couples into a different strip implant in parallel to the strip as sketched in the right drawing in figure 8.31. As expected, these effects are not observed for hits outside of the PA area. In figure 8.32 the same plot is shown for a **SensorST** as a reference and for **Sensor2MPA** which is heavily distorted due to the many bad strips. Nevertheless, the characteristic lines, caused by the layout of the PA in **SensorPA**, are not visible. This leads to the tentative conclusion, that the effect of crosstalk caused by the PA on double metal sensors is much less pronounced than for the **SensorPA** solution.

8.5. Beam Tests at CERN's SPS

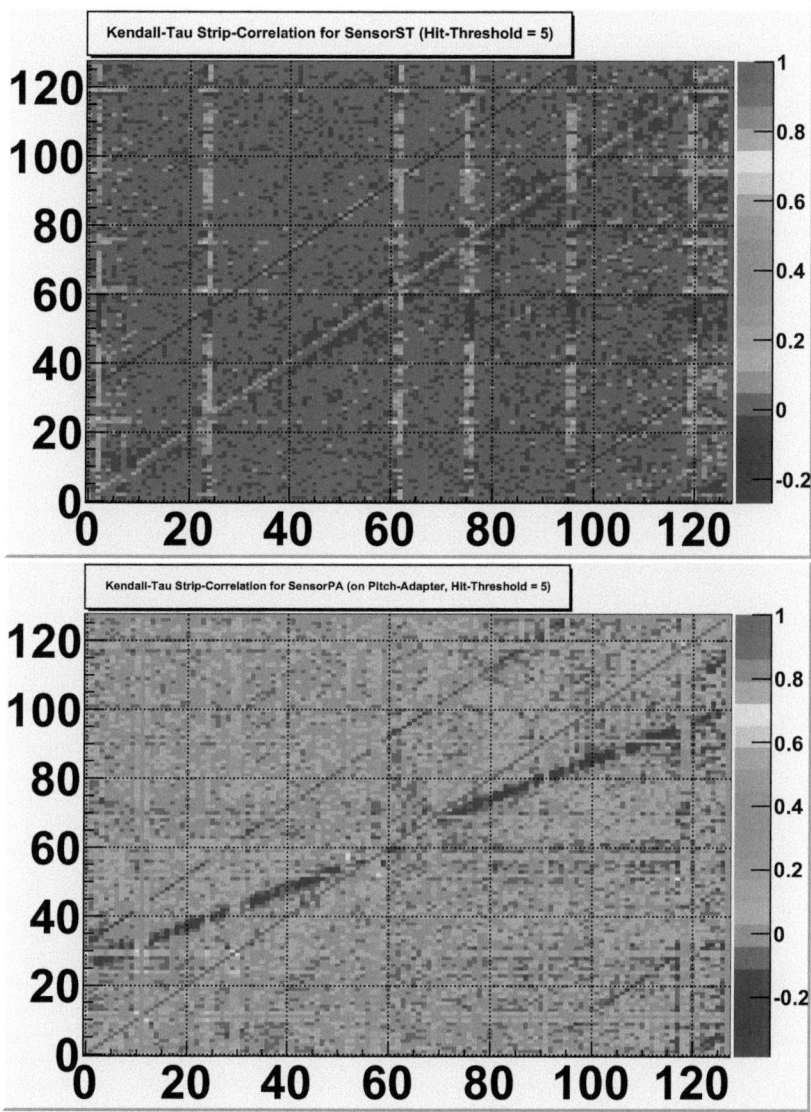

Figure 8.29.: Kendall tau rank correlation coefficient for all pairs of strips for a sensor without on-sensor pitch adapter (**SensorST**, top) and a sensor with an on-sensor pitch adapter in the first metal layer (**SensorPA**, bottom). The hits on SensorPA are from inside the pitch adapter area only. The horizontal structures are artifacts caused by bad strips.

8. Sensors with Integrated Pitch Adapters

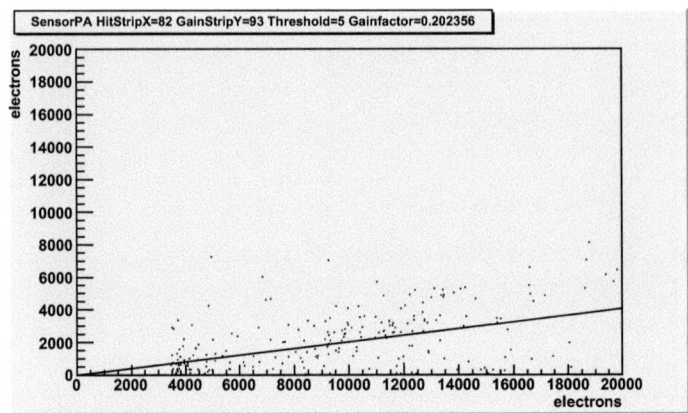

Figure 8.30.: Determination of the gainfactor for a single pair of strips. Strip 82 is hit by a particle and the signal measured is assigned along the X-axis. The signal measured on strip 93 at the same time is assigned along the Y-axis for each point. The slope of the line fit is the desired gainfactor for the pair 82-93.

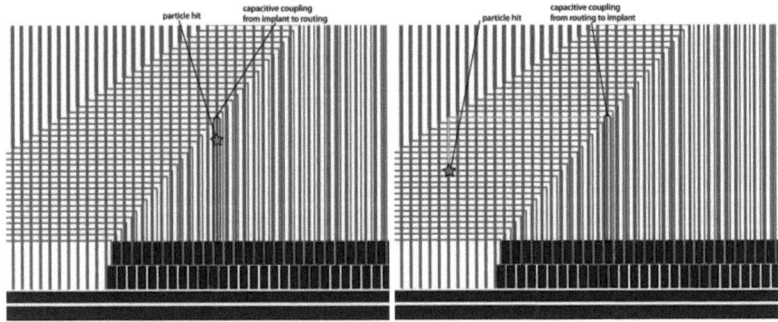

Figure 8.31.: A particle hits a strip where a routing line runs in parallel and near to the implant (left). The signal is directly coupled to the routing. For strips in the center, this is the more likely situation and the length of the parallel sections of the routing is longer. If the particle hits a strip without a parallel routing near to the implant, the coupling occurs between its routing line to an implant (right). For strips towards the edge of the sensor, this is the more likely situation, where the parallel sections of the routing are shorter.

8.5. Beam Tests at CERN's SPS

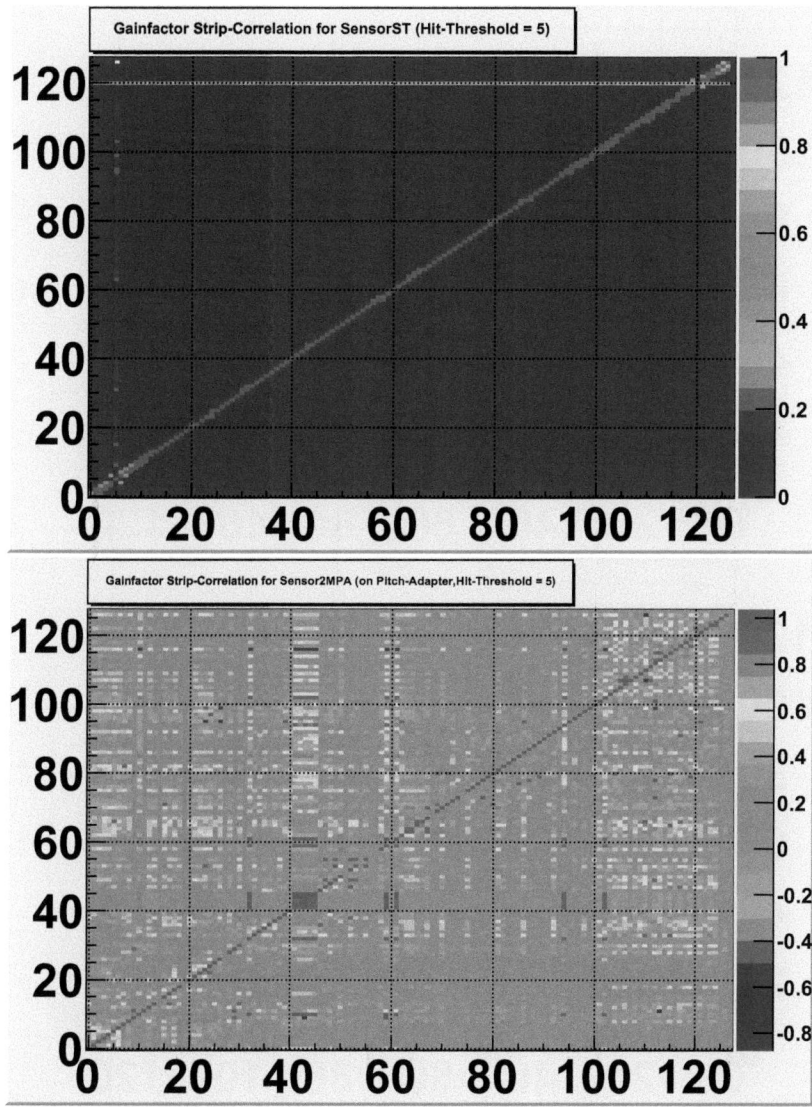

Figure 8.32.: Gainfactor for all pairs of strips where the strip hit by a particle (above 5 σ noise) is indicated on the X-axis and the other strips not hit at the same time are indicated on the Y-axis. The situation is shown for a sensor without an on-sensor pitch adapter (**SensorST**, top) and on a sensor with a pitch adapter on the second metal layer (**Sensor2MPA**, bottom). Due to the many bad channels, the full sensor seems to be saturated in the lower plot.

8. Sensors with Integrated Pitch Adapters

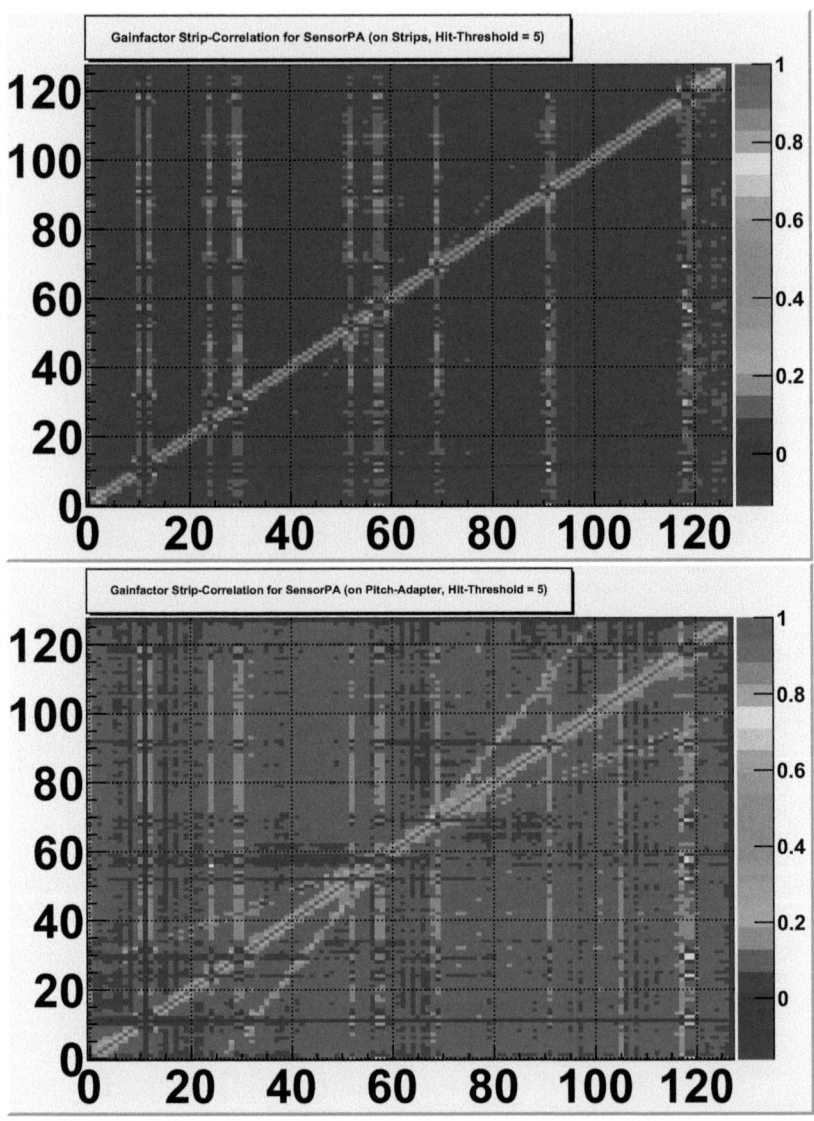

Figure 8.33.: Gainfactor for all pairs of strips where the strip hit by a particle (above 5 σ noise) is indicated on the X-axis and the other strips not hit at the same time are indicated on the Y-axis. The situation is shown for a sensor with a pitch adapter in the first metal layer (**SensorPA**, where hits are only located outside the pitch adapter area (top) or where the its are only located inside the pitch adapter area (bottom).

8.6. Summary

An novel concept of connecting readout chips to sensors has been presented in this chapter. It aims at integrating the functionality of a pitch adapter into the sensor itself. I have presented several methods of achieving this goal, like the integration in the first metal layer or adding a second metal layer to perform the routing. Very advanced concepts which have the readout chip bump bonded directly on top of the sensors have been introduced as well.

Several prototype sensor with different pitch adapter designs and integration methods have been produced with ITE Warsaw. The wafers also contained the set of test structures described in chapter 7, which were used to asses the process quality of the run. The results from the measurements of the test structures showed some significant deviations from the parameters specified by me and ITE Warsaw, like a thicker readout oxide and very low polysilicon resistivity. The same issues were also seen in the electrical characterisation of the sensors. A very serious issue is the mechanical instability of the oxide layers which produced many bad channels during the wire bonding of the sensor. All the mentioned problems were discussed with ITE Warsaw and possible solutions for a new run were identified.

The sensors were tested in a beam at the SPS at CERN. The interesting issues I investigated are signal loss and additional crosstalk introduced by the on-sensor pitch adapter. A signal reduction for sensors with a pitch adapter in the first metal layer can be clearly observed. While achieving a signal-to-noise ration of approximately 20 for hits outside of the pitch adapter area, the ratio can drop down to 10 inside the pitch adapter area. The signal loss is proportional to the length of the strip not covered by aluminium. Similar results have been obtained for the crosstalk investigations. Hits inside the pitch adapter area produce significant crosstalk where up to 50% of the signal couples into certain strips defined by the routing geometry.

These results show, that the **SensorPA** design with an on-sensor pitch adapter is a well working design for hits outside the area of the routing. Hits on the sensor within this area are subject to large signal loss and strong crosstalk. The **Sensor2MPA** design, were the routing is moved into a second metal layer, should reduce these issues significantly. Unfortunately, the low mechanical quality of the oxide layers has introduced many bad strips during the bonding process, making the results not entirely conclusive. A tentative interpretation of the results suggests, that the issues seen on **SensorPA** are very much mitigated in the double metal implementation. Further investigation with new sensors of the same design but an improved manufacturing process are needed and already envisaged for 2010.

9

Conclusion and Outlook

The upgrade of the LHC accelerator to the high luminosity sLHC will put strong demands on the tracker of the CMS experiment. The increase in particle interactions will require new radiation hard materials for the silicon strip sensors and a approach on sensor and module design to produce highly integrated detector modules.

The review of the current status of the RD50 research project in chapter 6 provided a number of promising candidates as bulk material for the middle to outer layers of the CMS Tracker upgrade. The most favorable materials in terms of radiation hardness today, are thin n-type or p-type MCz with n+ strips. Nevertheless, to conclude the selection process, a full study of all available options has to be done, where all samples are produced by a single high quality manufacturer, to mitigate the influence of the quality of the manufacturing process on the results. Such an order is now in production at Hamamatsu, Japan. The results and conclusions from a large irradiation campaign performed on the sensors and test structures will enable the final selection of a suitable material and production process.

To asses the quality of radiation hard silicon strip sensors, specialized test structures are required to measure all relevant parameters. The design of the structures needs to be flexible to adapt them to the different manufacturing processes available, while enabling the reliable measurement of all relevant values. Originating from the set of test structures used to monitor the production quality of the original strip sensors of the CMS Tracker, we have suggested

9. Conclusion and Outlook

several improvements to the existing structures. These structures have been manufactured by ITE Warsaw in two production runs and electrically characterised at the HEPHY in Vienna. All improvements were validated successfully including a new structure to measure the oxide thickness between two metal layers.

Several advanced sensors have been designed and later produced at ITE Warsaw. The sensor layouts were created with the aim of integrating the functionality of an external pitch adapter onto the sensor. This concept allows the construction of light strip detector modules with a very high density of readout channels. After the careful electric characterisation of the sensors, they have been field tested in a beam test experiment. The results show, that the concept of an on-sensor pitch adapter in the first metal layer is a cost effective solution with reduced performance in the area of the pitch adapter. These shortcomings should be mitigated by the more cost-intensive double metal solution. Unfortunately, the measurements on these double metal sensors with on-sensor pitch adapter are distorted by a large number of bad channels which are caused by a low quality oxide. A new run of double metal sensors with on-sensor pitch adapters at ITE Warsaw is already in production as of mid 2010. It will feature an improved manufacturing process to overcome the problems of the low oxide quality.

Ultimately, the double metal sensors with on-sensor pitch adapters will enable the construction of the highly integrated detector modules needed for the CMS Tracker upgrade.

Bibliography

[1] G. Lutz. *Semiconductor Radiation Detectors: Device Physics.* Springer, Berlin, 2007. ISBN: 3540716785. 7, 24, 33, 38

[2] S. M. Sze. *Physics of Semiconductor Devices.* Wiley & Sons, 1981. ISBN: 0471056618. 14, 15, 25, 30, 57

[3] W.-M. Yao et al. *Review of Particle Physics – Passage of particles through matter.* Journal of Physics G, 33, 2006. 18, 19

[4] E. Barberis et al. *Capacitances in silicon microstrip detectors.* Nuclear Inst. and Methods in Physics Research, A, 342(1): 90–95, 1994. 28, 29

[5] G. L. Ì. A. Vasilescu. *Notes on the fluence normalisation based on the NIEL scaling hypothesis.* ROSE/TN/2000- 02, 2000. 38

[6] G. Lindström. *Radiation damage in silicon detectors.* Nuclear Inst. and Methods in Physics Research, A, 512(1-2): 30–43, 2003. 40

[7] C. Bozzi et al. *Test Results on Heavily Irradiated Silicon Detectors for the CMS Experiment at LHC.* IEEE Transactions on Nuclear Science, 47(6), December 2000. 42, 43, 44, 45

[8] RD48 Collaboration. *3rd RD48 Status Report.* CERN/LHCC 2000-009, 2000. 42, 114

[9] J. Kemmer. *Fabrication of low noise silicon radiation detectors by the planar process.* Nuclear Inst. and Methods, 169(3): 499–502, 1980. 47

[10] W.-M. Yao et al. *Review of Particle Physics.* Journal of Physics G, 33: 258–270, 2006. 48, 50

[11] D. Landau. *On the Energy Loss of Fast Particles by Ionization.* In Collected Papers of L. D. Landau. Pergamon Press, Oxford, 1965. 49

Bibliography

[12] H. Bichsel. *Straggling in thin silicon detectors.* Reviews of Modern Physics, 60(3): 664–669, 1988. 49

[13] S. Ramo. *Currents Induced by Electron Motion.* Proceedings of the I.R.E. 27, 27: 584–585, 1939. see also http://www-physics.lbl.gov/~spieler/physics_198_notes_1999/PDF/IX-1-Signal.pdf. 49

[14] F. Hartmann. *Evolution of Silicon Sensor Technology in Particle Physics*, volume 231 of Springer Tracts in Modern Physics. Springer, 2009. 52, 54

[15] R. Richter et al. *Strip detector design for ATLAS and HERA-B using two-dimensional device simulation.* Nuclear Inst. and Methods in Physics Research, A, 377(2-3): 412–421, 1996. 58

[16] U. Hilleringmann. *Silizium-Halbleitertechnologie.* Teubner, Wiesbaden, 2004. 60, 61

[17] ATLAS Colaboration. *ATLAS : letter of intent for a general-purpose pp experiment at the large hadron collider at CERN.* CERN-LHCC-92-004; LHCC-I-2, 1992. 76

[18] CMS Collaboration. *CMS Technical Design Report – The Tracker Project.* CERN/LHCC 98-6, 1998. 79, 80, 107, 108

[19] CMS Collaboration. *CMS Technical Design Report – The Electromagnetic Calorimeter.* CERN/LHCC 97-33, 1997. 82

[20] CMS Collaboration. *CMS Technical Design Report – The Hadronic Calorimeter.* CERN/LHCC 97-31, 1997. 83

[21] CMS Collaboration. *CMS Technical Design Report – The Muon Project.* CERN/LHCC 97-32, 1997. 84

[22] CMS Collaboration. *CMS Technical Design Report – The Trigger Systems.* CERN/LHCC 2000 - 38, 2000. 85

[23] S. Braibant et al. *Investigation of design parameters for radiation hard silicon microstrip detectors.* Nuclear Inst. and Methods in Physics Research, A, 485(3): 343–361, 2002. 88

[24] CMS Tracker Collaboration. *Supply of Silicon Micro-Strip Sensors for The CMS Silicon Strip Tracker.* IT-2777/EP/CMS, 2000. 88, 89

Bibliography

[25] J. Agram et al. *The silicon sensors for the Compact Muon Solenoid tracker – design and qualification procedure*. Nuclear Inst. and Methods in Physics Research, A, 517(1-3): 77–93, 2004. 88, 90

[26] RD48 Collaboration, *RD48 - Research and development On Silicon for future Experiments*. URL: www.cern.ch/rd48. 88

[27] A. Furgeri. *Quality Assurance and Irradiation Studies on CMS Silicon Strip Sensors*. PhD thesis, Institut für Experimentelle Kernphysik, Universität Karlsruhe, 2005. IEKP-KA/2005-1. 89

[28] T. Bergauer. *Process Quality Control of Silicon Strip Detectors for the CMS Tracker*. Master's thesis, TU Wien, 2004. 92, 125, 126, 159

[29] M. French et al. *Design and results from the APV25, a deep sub-micron CMOS front-end chip for the CMS Tracker*. Nuclear Inst. and Methods in Physics Research, A, 466(2): 359–365, 2001. 92

[30] F. Gianotti et al. *Physics potential and experimental challenges of the LHC luminosity upgrade*. European Physical Journal, c39: 293–333, 2005. CERN-TH/2002-078. 100, 108, 109

[31] Y. Papaphilippou and F. Zimmermann. *Estimates of diffusion due to long-range beam-beam collisions*. Phys. Rev. ST Accel. Beams 5, 5(074001), 2002. 101

[32] LHC-LUMI-06 Workshop, *Parameter list compiled from various sources as reference for the Workshop*. URL: http://care-hhh.web.cern.ch/CARE-HHH/LUMI-06/. 103

[33] CMS Collaboration. *Addendum to the CMS Tracker TDR*. CERN/LHCC 2000-016, February 2000. 108

[34] CMS Collaboration. *Technical Design Report – Data Acquisition and High-Level Trigger*. CERN/LHCC 2006-001, 2002. 110

[35] CMS Collaboration. *Technical Design Report – Detector Performance and Software*. CERN/LHCC 2006-001, 2006. 110, 111

[36] G. Hall et al. *Prototyping stacked modules for the L1 track trigger*. CMS Upgrade R&D Proposal. 112

Bibliography

[37] RD50 Collaboration, *RD50 - Radiation hard semiconductor devices for very high luminosity colliders*. URL: http://www.cern.ch/rd50. 114, 132

[38] P. P. Allport, G. Casse, and A. Greenall. *Radiation tolerance of oxygenated n-strip readout detectors*. Nuclear Inst. and Methods in Physics Research, A, 438(2-3): 429–432, 1999. 114

[39] RD48 Collaboration. *Radiation hard silicon detectors - developments by the RD48 (ROSE) collaboration*. Nuclear Inst. and Methods in Physics Research, A, 466(2): 308–326, 2001. 114

[40] Z. Lia et al. *Gamma radiation induced space charge sign inversion and re-inversion in p-type MCZ Si detectors and in proton-irradiated n-type MCZ Si detectors*. Nuclear Inst. and Methods in Physics Research, A, 552(1-2): 34–42, 2005. 114

[41] E. Fretwurst et al. *Survey of recent radiation damage studies at Hamburg.* presented at the Third RD50 Workshop, CERN, Geneva, Switzerland, November 2003. 114

[42] D. Creanza et al. *Comparison of the radiation hardness of Magnetic Czochralski and Epitaxial silicon substrates after 26 MeV proton and reactor neutron irradiation*. Nuclear Inst. and Methods in Physics Research, A, 579(2): 608–613, 2007. 114

[43] G. Pellegrini et al. *Annealing Studies of magnetic Czochralski silicon radiation detectors.* Nuclear Inst. and Methods in Physics Research, A, 552(1-2): 27–33, 2005. 114

[44] J. Härkönen et al. *Radiation hardness of Czochralski silicon, Float Zone silicon and oxygenated Float Zone silicon studied by low energy protons*. Nuclear Inst. and Methods in Physics Research, A, 518(1-2): 346–348, 2004. 114

[45] B. Dezilliea et al. *Radiation hardness of silicon detectors manufactured on wafers from various sources*. Nuclear Inst. and Methods in Physics Research, A, 388(3): 314–317, 1997. 115

[46] G. Kramberger. *Charge collection measurements on MICRON RD50 detectors.* presented at the ATLAS Tracker Upgrade Workshop, Valencia, Italy, December 2007. 115, 116, 119, 132

[47] A. Bocci et al. *The Powering Scheme of the CMS Silicon Strip Tracker*. In Proceedings of the 10th Workshop on Electronics for LHC Experiments, Boston, 2004. CERN. 115

[48] RD50 Collaboration. *RD50 Status Report 2006*. CERN/LHCC 2007-005, 2006. 117

Bibliography

[49] L. Andricek et al. *Processing of ultra thin silicon sensors for future e+e- linear collider experiments*. Nuclear Science Symposium Conference Record, 2003 IEEE, 3: 1655–1658, 2003. 118

[50] P. P. Allport, A. Affolder, and C. G. *Charge collection efficiency measurement for segmented silicon detectors irradiated to* $1 \times 10^{16} \, n \, cm^{-2}$. IEEE Trans. Nucl. Sci., 55(3): 1695–1699, June 2008. 119, 120, 121, 122

[51] J. Lange. *Charge multiplication in radiation-damaged epitaxial silicon detectors*. 12th Vienna Conference on Instrumentation, 2010. 121, 123

[52] T. Ohsugi et al. *Microdischarges of AC-coupled silicon strip sensors*. Nuclear Inst. and Methods in Physics Research, A, 342(1): 22–26, 1994. 121

[53] M. Dragicevic. *Quality Assurance and Performance Tests of Silicon Detector Modules for the CMS/Tracker*. Master's thesis, TU Wien, May 2005. 123

[54] M. Manneli. *R&D for Thin Single-Sided Sensors with HPK*. CMS Upgrade R&D Proposal. 124, 142

[55] M. Dragicevic et al. *Results from a first production of enhanced Silicon Sensor Test Structures produced by ITE Warsaw*. Nuclear Inst. and Methods in Physics Research, A, 1(1): 86–88, 2009. 132

[56] C. Broennimann et al. *Development of an Indium bump bond process for silicon pixel detectors at PSI*. Nuclear Inst. and Methods in Physics Research, A, 565(1): 303–308, 2006. 146

[57] A. Klumpp et al. *Vertical System Integration Technology for High Speed Applications by Using Inter-Chip Vias and Solid-Liquid Interdiffusion Bonding*. Jpn. J. Appl. Phys, 43: L829 – L830, 2004. 146, 148

[58] C. Irmler. *Upgrade Studies for the Belle Silicon Vertex Detector*. TU Wien, August 2008. 147

[59] T. Bergauer. *Design, Construction and Commissioning of the CMS Tracker at CERN and Proposed Improvements for Detectors at the Future International Linear Collider*. PhD thesis, TU Wien, 2008. 159

Bibliography

[60] M. Friedl, C. Irmler, and P. M. *Occupancy reduction in silicon strip detectors with the APV25 chip.* Nuclear Inst. and Methods in Physics Research, A, 569(1): 92–97, December 2006. 169

[61] T.Haas. *A PixelTelescope for Detector R&D for an International Linear Collider.* Nuclear Inst. and Methods in Physics Research, A, 569: 53–56, 2006. 169

[62] M. Raymond et al. *The CMS Tracker APV25 0.25 µm CMOS Readout Chip.* 6th Workshop on Electronics for LHC Experiments, Cracow, Poland, 2000. 171

[63] L. Jones, *APV25-S1 Users Guide Version 2.2.* URL: http://www.te.rl.ac.uk/med/projects/High_Energy_Physics/CMS/APV25-S1/pdf/User_Guide_2.2.pdf. 171

[64] M. Raymond et al. *Final Results from the APV25 Production Wafer Testing.* In Proceedings of the 11th Workshop on Electronics for LHC and Future Experiments, pages 453–457, Heidelberg, September 2005. 171

[65] M. Kendall. *A New Measure of Rank Correlation.* Biometrika, 30(1-2): 81–93, 1938. 175

Die VDM Verlagsservicegesellschaft sucht für wissenschaftliche Verlage abgeschlossene und herausragende

Dissertationen, Habilitationen, Diplomarbeiten, Master Theses, Magisterarbeiten usw.

für die kostenlose Publikation als Fachbuch.

Sie verfügen über eine Arbeit, die hohen inhaltlichen und formalen Ansprüchen genügt, und haben Interesse an einer honorarvergüteten Publikation?

Dann senden Sie bitte erste Informationen über sich und Ihre Arbeit per Email an *info@vdm-vsg.de*.

Sie erhalten kurzfristig unser Feedback!

VDM Verlagsservicegesellschaft mbH
Dudweiler Landstr. 99
D - 66123 Saarbrücken
www.vdm-vsg.de

Telefon +49 681 3720 174
Fax +49 681 3720 1749

Die VDM Verlagsservicegesellschaft mbH vertritt

Printed by Books on Demand GmbH, Norderstedt / Germany